D1300877

COMPREHENSIVE CHEMICAL KINETICS

COMPREHENSIVE

CHEMICAL KINETICS

EDITED BY

R.G. COMPTON

M.A., D. Phil. (Oxon.)

University Lecturer in Physical Chemistry
and Fellow, St. John's College, Oxford

VOLUME 33

CATASTROPHE THEORY

A. OKNIŃSKI

Warsaw University of Technology

ELSEVIER

AMSTERDAM–LONDON–NEW YORK–TOKYO

PWN – POLISH SCIENTIFIC PUBLISHERS

WARSZAWA

1992

Translated by *Andrzej Przyjazny* from the Polish edition
Teoria katastrof w chemii, published by Państwowe Wydawnictwo Naukowe, Warszawa 1990

Distribution of this book is being handled by the following publishers:

For the USA and Canada
ELSEVIER SCIENCE PUBLISHING CO., INC.
655 Avenue of the Americas, New York, NY 10010

For Albania, Bulgaria, Cuba, Czecho-Slovakia, Hungary, Korean People's Democratic
Republic, Mongolia, People's Republic of China, Poland, Romania, the USSR, Vietnam and
Yugoslavia
POLISH SCIENTIFIC PUBLISHERS PWN LTD.
Miodowa 10, 00-251 Warszawa

For all remaining areas
ELSEVIER SCIENCE PUBLISHERS B.V.
Sara Burgerhartstraat 25
P.O. Box 211, 1000 AE Amsterdam, The Netherlands

Library of Congress Cataloging-in-Publication Data

Okniński Andrzej.
　　[Teoria katastrof w chemii. English]
　　Catastrophe theory / A. Okniński; [translated from the Polish by
　　Andrzej Przyjazny].
　　p.cm. -- (Comprehensive chemical kinetics; v. 33)
　　Translation of: Teoria katastrof w chemii.
　　Includes bibliographical references and index.
　　ISBN 0-444-98742-8
　　1. Chemical reactions, Conditions and laws of. 2. Catastrophies
　　(Mathematics) I. Title. II. Series.
QD501. B242 vol. 33
541. 3′9 s--dc20
[541. 3′9]　　　　　　　　　　　　　　　　　　　　　　　　　　　91-28894
　　　　　　　　　　　　　　　　　　　　　　　　　　　　　　　　　CIP

QD501
B242
vol 33

ISBN 0-444-98742-8

Copyright © by Polish Scientific Publishers PWN Ltd. — Warszawa 1992

Printed in Poland

COMPREHENSIVE CHEMICAL KINETICS

ADVISORY BOARD

Volumes in the Series

Contributor to Volume 33

All chapters in this volume have been written by

A. OKNIŃSKI   Institute of Organic Chemistry,
             Warsaw University of Technology,
             Warsaw, Poland

Preface

Volume 33 of Comprehensive Chemical Kinetics is concerned with Catastrophe Theory and Chemistry. The book embraces both fundamentals and applications. The author's detailed view of the objectives and scope of the book are summarised in his preface which follows.

Oxford   *R.G. Compton*
August 1991

Author's Preface

Recently, visible progress has been achieved in the theoretical and experimental investigations of non-linear phenomena. Among vital achievements in this area, the theory of solitons, strange attractors, the theory of fractals, chemical reactions of complex dynamics should be mentioned.

Investigations of the chemical reactions, in which concentration oscillations, formation of non-homogeneous complex structures from homogeneous solution and the occurrence of chaotic dynamics have been observed, seem to be particularly promising. The question concerning the nature of couplings between chemical reactions is considered to be crucial in our anxiety to understand life.

One of the theories of non-linear phenomena, recently created and being rapidly developed, is catastrophe theory. Catastrophe theory deals with the non-linear phenomena in which a continuous change in control parameters results in a discontinuous alteration of a quantity characterizing the examined system. It is well suited for the investigation of non-linear equations of chemical kinetics, describing chemical reactions.

The scope of this book on catastrophe theory is chemistry. Consequently, all calculations and derivations necessary to understand the material have been presented in the book. The calculus of catastrophe theory introduced in the book has been applied to chemical kinetic equations. Chemical reactions without diffusion are classified from the standpoint of catastrophe theory and the most recent theoretical results for the reactions with diffusion are presented. The connections between various domains of physics and chemistry dealing with nonlinear phenomena are also shown and the progress which has been recently achieved in catastrophe theory is presented.

Andrzej Okniński

Contents

Introduction

In the second half of the 20th century a distinct change in the way of scientific thinking took place. The change was associated with an appearance of several novel concepts in science, involving a comprehensive treatment of the problems considered, in contrast with the method of analysis — dividing the problem into parts, deriving from Descartes. Each of these new theories has been initiated by one man. Ludwig von Bertalanffy created the theory of systems, Norbert Wiener — cybernetics, Herman Haken initiated synergetics (theory of cooperative phenomena), and a decisive incentive to develop catastrophe theory has been given by René Thom. All these theories are characterized by a general approach to the examined phenomena and emphasize, not neglect, their non-linear features. Previously, there has been a distinct trend to disregard non-linearity and to search for a description of the investigated phenomenon by way of linearization. Such a trend could have resulted from the lack of mathematical methods permitting an effective investigation of non-linear equations. The first great non-linear theories were hydrodynamics and general relativity theory. The creation of these theories have forced mathematicians and physicists to study non-linear problems. As a result, a visible progress in the theory of non-linear differential equations took place soon after. Among significant achievements in the area of non-linear investigations, the theory of solitons, strange attractors and fractal theory should be mentioned.

This also gave rise to catastrophe theory, whose philosophical, mathematic and natural genesis will be elucidated in Chapter 1. It should be pointed out that mentioned theories, similarly to other methods of examination of non-linear phenomena (for example synergetics or bifurcation theory closely related to catastrophe theory), have numerous common points. Consequently, some researchers of non-linear phenomena expect the future development future of a general theory allowing the investigation of all aspects of non-linear processes.

It should be stated at this point what is meant by the term catastrophe and what the relation can be between catastrophe theory and chemical reactions. We shall refer to a catastrophe as the phenomenon consisting in the loss of stability by a formerly stable state of the system followed by a rapid transition to another state of the system, stable under the new conditions. The loss of stability results from a change of parameters determining the state of the system; the changes in parameters, in contrast with the transition to a new state which is generally abrupt (hence the term catastrophe), are slow and continuous.

Until recently, the domain of catastrophe theory was mainly physics. The situation has changed after recent discoveries in chemical reactions of many phenomena associated with a loss of stability, inspired by Prigogine theory of systems far from thermodynamical equilibrium. Suffice it to mention the discovery of spatial and oscillating structures in homogeneous systems of the generation of turbulent states in flow reactors. These results are of vital importance in the theory of chemical reactors and in the investigation of living organisms, forming the basis for theoretical elucidation of tissue diversification or an explanation of the action of biological clocks. The ease of experimental generation of simple chemical systems having very complex dynamics makes them a very convenient field of applications of non-linear theories, such as synergetics or catastrophe theory.

The catastrophes appearing in chemical systems should not be considered in separation from sometimes very similar phenomena occurring in physics or biology. Examples from the domains of mathematics, physics, physical chemistry and even biology allow, on the one hand, to place the concepts and methods of catastrophe theory in a broader context and, on the other, to have a more general point of view on chemical reactions.

The growing complexity of phenomena observed in chemistry means that they are increasingly more difficult to examine without resorting to rather complicated mathematical theories, to which catastrophe theory belongs. Difficult problems, however, are usually interesting.

The origin of catastrophe theory is discussed in the first chapter of the book. Examples of various catastrophes occurring in physics, chemistry and biology are given. Chapter 2 contains fundamental knowledge on elementary catastrophe theory, being a starting-point to more general considerations. Some additional information is provided in the Appendix. In Chapter 3, applications of elementary catastrophe theory to the description of some static catastrophes of physics, physical chemistry and biology described in Chapter 1 are presented (chemical catastrophes, the main topic of the book, will be analysed in Chapter 6). Fundamentals of chemical kinetics are discussed in Chapter 4, while in Chapter 5 the methods of analysis of chemical kinetic equations using the methods of elementary catastrophe theory and generalized catastrophe theory are reviewed. The Appendix to this chapter contains supplementary mathematical knowledge.

Chapter 6 provides applications of the formalism introduced in Chapters 2 and 5 to chemical reactions and a brief discussion of the applications of catastrophe theory to the analysis electron denisty distribution in chemical molecules and their changes in the course of a reaction.

Chapter 1

Origin of Catastrophe Theory

Intellectual movements which have led to the development of catastrophe theory have their origin in philosophy, mathematics and the experimental sciences. The evolution of philosophical views, the progress in the area of non-linear mathematical methods and the development of experimental methods related to the investigation of stability of processes have created favourable conditions for the appearance of catastrophe theory.

1.1 PHILOSOPHY

Program assumptions of catastrophe theory are associated with certain philosophical concepts. The philosophical context of catastrophe theory is essential, as the description of reality in this theory is deliberately limited to some forms of things and phenomena only. More specifically, the theory is an attempt at the description of only such phenomena whose form is resistant to perturbations, that is, structurally stable. In addition, catastrophe theory describes changes in the form taking place as a result of continuous variations of control parameters, emphasizing qualitative, structurally stable changes in the form.

A fundamental concept in catastrophe theory is the structurally stable form (this can be a form of thing or process). Such a concept is close to the Platonic concept of idea or to the Aristotelian concept of form. In the Plato system the basic concept is an invariable and absolute idea, while things are imitations of the idea. In the Aristotle system it is assumed that matter and form exist combined into entirety. Form is more important (being an equivalent of the Plato's idea), since only form, an essential and permanent component of things, is cognizable. Hence, structural stability of forms in the Thom theory corresponds, on the one hand, to the invariable and indestructible ideas of Plato and, on the other, to the forms of Aristotle.

The concept of a structurally stable form is also related to a philosophical and experimental problem of the recognition of forms. Note that in general we have no troubles with the recognition of things, objects and phenomena despite tha fact that the object or process under investigation frequently occurs in a deformed shape. For example, a tree does not cease to be a tree after falling of leaves or even after cutting off a number of branches. Apparently, the possibility of identification of form is associated with its structural stability. In this respect catastrophe theory seems to be closely related to views of the twentieth century "Gestalttheorie" (theory of shape, form), whose outstanding representative was W. Köhler, the author of pioneering studies on the learning of apes, and M. Wertheimer. Philosophers belonging to this school connected the possibility of identification, recognition of a problem with thought directing, called by them centering, by most significant features of the perceived situation. According to Wertheimer, the centering forces are inherent to the problem situation, inevitably directing thoughts to the essence of the situation. Classification of forms in catastrophe theory, pertaining to spatial sets, is close to this idea, the centering force in catastrophe theory being structural stability of a spatial form, for example of a tree.

The second important contact point of catastrophe theory with the philosophical problems is a description of the qualitative change of a form, called a catastrophe, occurring as a result of continuous variations of control parameters. The ideas claiming that the qualitative change of a system may be considered as an effect of slow quantitative changes in some of its parameters may be related to Hegel and Engels. Considering a phase transition in the water–vapour or water–ice system Hegel believed that "initially water temperature has no effect on its liquid state; later, however, on increasing or decreasing a temperature of liquid water there comes a moment when cohesion undergoes a change and water turns in one case into vapour and in the second case − into ice". In the example given by Hegel control parameters are temperature of water and pressure (constant in the above example), while a state parameter (to be measured) is water density. Hegel included the essence of his considerations in the law of conversion of quantitative into qualitative changes. The law has been further developed by Engels who wrote: "Each liquid has at a given pressure its melting and boiling point (...), finally each gas has its critical point at which it is liquified under suitable pressure and cooling. In other words: so-called physical 'constants' are usually nothing else but the names of nodal points at which a quantitative change, increase or decrease of motions, causes in the state of a given body a qualitative change, hence at which quantity turns to quality". In catastrophe theory, to a node at which the transition of quantity into quality takes place corresponds the catastrophe set, in which the state of the system is sensitive (susceptible to disturbances). A catastrophe set divides the states of a system, the transition between which signifies a catastrophe.

1.2 MATHEMATICS

In mathematics, the notion about superiority of quantitative over qualitative methods has prevailed for a long time, beginning from the times of Newtonian mechanics. At that time, this resulted from the development of new mathematical methods, differential and integral calculus, enabling an effective solution of physical problems. However, a converse trend has begun to slowly appear. Firstly, doubts concerning superiority of quantitative over qualitative methods have been raised. Let us consider an example presented by René Thom. Suppose that an investigated phenomenon can be described by an experimental curve $g(x)$, while curves $g_1(x)$ and $g_2(x)$ correspond to two theories T_1 and T_2, respectively (Fig. 1).

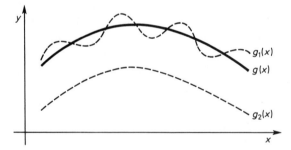

Fig. 1. Experimental function $g(x)$ and two theoretical functions $g_1(x)$ and $g_2(x)$

Apparently, the theoretical curve g_1 better agrees with the experimental curve g in quantitative terms. On the other hand, the curves g_2 and g are substantially more similar qualitatively (have the same form). Thom believes that a theoretician should accept the theory T_2, since it should account more realisticall for the mechanisms underlying the examined phenomenon than the theory T_1.

Secondly, the computational effectiveness of quantitative methods requires restriction to analytic (continuous) functions. The restriction to a description of phenomena by means of continuous functions implies resignation from the description of discontinuous phenomena. And discontinuous phenomena, as we shall demonstrate in the following section, do occur in nature. We shall restrict ourselves here to the example of breaking of the crest of a sea wave, given after Thom. Seemingly, one should not a priori choose calculation as the only solution and deliberately give up the description of discontinuous phenomena. Thus, the need for application of qualitative methods in mathematics and physics is obvious.

Imperfections of computational methods do not end at that. A computational approach has a drawback, discovered relatively recently — the examples of occurrence of insolvable mathematical problems are becoming increasingly more common; for example, the proofs of Abel and Galois on non-existence of such a general formula for the roots of a polynomial of order higher than fourth which would contain only four arithmetic operations and extraction of a root. Apparently, the roots of any polynomial are calculable — they can be computed with an arbitrary accuracy. However, computational problems have appeared for which the solutions do not exist. Turing and Post, inspired to some extent by the work of Gödel, provided the examples of mathematical problems for which no computational solution (algorithm) can be given. Since then, it has become evident that there is a whole class of incalculable problems. As an example we shall consider a computer problem reported by Turing. The problem involves writing a program (giving a computational algorithm) permitting calculation whether a given program being tested will halt after some time or it will loop (acting to infinity). It turns out that the above computational problem is insoluble. The tenth Hilbert's problem, consisting in provision of a computational procedure to establish whether any equation with whole coefficients, e.g. $x^n + y^n = z^n$, where n is any integer, has a solution in integers, was found to be another insolvable computational problem.

Qualitative methods have not long been developed due to a lack of precise determination of the function form. However, owing to a progress in the field of differential geometry, a precise definition of the differential type (form) of a function has become possible (Chapter 2). The appearance of a good definition of the function form has enabled Thom and other mathematicians to examine changes in the function form in relation to variations of parameters on which the investigated function depended.

Vitally important for catastrophe theory is the notion of structural stability which resulted from the studies on stability of the solar (planetary system), initiated by Laplace and continued by Poincaré and Arnol'd. The notion of structural stability was introduced in 1937 by Andronov and Pontryagin. They considered a system to be structurally stable if its dynamics were not qualitatively changed by a perturbation. Let us consider an example illustrating the notion of structural stability

$$\dot{x} = ax - y \tag{1.1a}$$

$$\dot{y} = x + ay \tag{1.1b}$$

where a is a parameter (a dot denotes the derivative with respect to time t). The solutions to the system (1) are of the form $x = x_0 e^{rt}$, $y = y_0 e^{rt}$. Substitution of these expressions to equation (1.1) yields

$$r = a \pm i \; (i^2 = -1) \tag{1.2}$$

The solutions to the system (1.1) may be plotted in the (x, y) plane, the trajectories of the system, $x(t)$, $y(t)$, depending on the parameter a (Fig. 2).

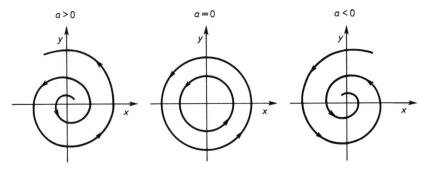

Fig. 2. Phase portrait for system (1) for $a > 0$, $a = 0$, $a < 0$.

It is evident that the system (1.1) for $a < 0$ (or $a > 0$) is structurally stable with respect to changes in the parameter a: a small change of a (not altering its sign) does not change the form of a trajectory on the phase plane $[x(t), y(t)]$. On the other hand, the system (1.1) for $a = 0$ is not structurally stable — an arbitrarily small perturbation of the parameter a change the form of the system trajectory.

The above analysis is characteristic of catastrophe theory. It can be said that the research program of this theory involves a qualitative investigation of types of solutions of equations describing some phenomena and physical processes, depending on parameters occurring in these equations. Using this type of analysis it can be established, for example, that the system under study may occur in states in which it is sensitive to arbitrarily small perturbations [see system (1)]. Futhermore, it may be proved that such a property of the system is structurally stable — it does not vanish under perturbation of the investigated system.

We shall now report after Gilmore the classification of parameter dependent equations, which may appear in the description of a phenomenon or process. Consider a possibly general system of equations (F_i are certain known functions):

$$F_i\{\boldsymbol{\psi}; \mathbf{c}; t, \mathrm{d}\boldsymbol{\psi}/\mathrm{d}t, \mathrm{d}^2\boldsymbol{\psi}/\mathrm{d}t^2, ...; \mathbf{x}, \mathrm{d}\boldsymbol{\psi}/\mathrm{d}x_1, \mathrm{d}^2\boldsymbol{\psi}/\mathrm{d}x_1\mathrm{d}x_m, ...$$
$$\int f(\boldsymbol{\psi}; \mathbf{c}; ...)\mathrm{d}x_1; ...\} = 0$$
$$i = 1, ..., N, \qquad l, m = 1, ..., n \tag{1.3a}$$

in which $\boldsymbol{\psi}$ are the so-called state functions (describing the state of the system), $\mathbf{c}$ are certain parameters determining experimental conditions (these are so-called control parameters), t stands for time, $\mathbf{x}$ is the coordinate vector:

$$\boldsymbol{\psi} = (\psi_1, ..., \psi_N), \qquad \mathbf{c} = (c_1, ..., c_k), \qquad \mathbf{x} = (x_1, ..., x_n) \tag{1.3b}$$

The problem of determination of the qualitative dependence of ψ_j on the parameters $\mathbf{c}$ is too difficult in the case of such a general equation. To be able to partly accomplish the research program of catastrophe theory, the class of the examined equations should be restricted. The system of equations

$$F_i(\boldsymbol{\psi}; \mathbf{c}; t, \mathrm{d}\boldsymbol{\psi}/\mathrm{d}t) = 0, \qquad i = 1, ..., N \tag{1.4}$$

in which the functions F_i depend explicitly on the time t still is an excessively complicated system. In a special case

$$F_i = \mathrm{d}\psi_i/\mathrm{d}t - f_i(\boldsymbol{\psi}; \mathbf{c}; t) = 0, \qquad i = 1, ..., N \tag{1.5}$$

we have to do with a so-called dynamical system. The functions f_i, having the sense of force, depend in equations (1.5) explicitly on time; such dynamical systems are called nonautonomous. For systems of this type only a few rigorous results have so far been obtained. On the other hand, in the case of autonomous dynamical systems (f_is do not depend explicitly on time), to which belong for example chemical kinetic equations,

$$F_i = \mathrm{d}\psi_i/\mathrm{d}t - f_i(\boldsymbol{\psi}; \mathbf{c}) = 0, \qquad i = 1, ..., N \tag{1.6}$$

there is a number of rigorous results obtained by means of the methods of catastrophe theory [equations of the form (1.6) are briefly discussed at the end of this section]. In the case when all forces f_i are potential, that is, when they may be written in the form

$$f_i = -\partial V(\boldsymbol{\psi}; \mathbf{c})/\partial \psi_i, \quad i = 1, ..., N \tag{1.7}$$

where V is the so-called potential function, the system (1.6) becomes a so-called gradient system

$$F_i = \mathrm{d}\psi_i/\mathrm{d}t + \partial V(\boldsymbol{\psi}; \mathbf{c})/\partial \psi_i = 0, \quad i = 1, ..., N \tag{1.8}$$

Numerous rigorous results have been obtained for gradient systems. One natural method of investigation of gradient systems is elementary catastrophe theory: the field of catastrophe theory dealing with an examination of gradient systems. In the case of the gradient system of equations (1.8), properties of a stationary state, that is the state invariant with time, may be readily studied

$$\mathrm{d}\psi_i/\mathrm{d}t = 0, \quad i = 1, ..., N \tag{1.9}$$

It follows from equations (1.8) and (1.9) that in the stationary state of a gradient system the condition

$$\partial V(\boldsymbol{\psi}; \mathbf{c})/\partial \psi_i = 0, \quad i = 1, ..., N \tag{1.10}$$

is satisfied which implies that the function V has a critical point. The critical point of a potential function corresponds to the so-called singular point of equations (1.8) — at this point the derivatives $\partial \psi_i/\partial \psi_j$ cannot be determined.

The problem of dependence of the type of stationary points and their stability on control parameters $\mathbf{c}$ is thus reduced for systems (1.8) to the investigation of a dependence of the type of critical points of a potential function V and their stability on these parameters. The above mentioned problems are, as already mentioned, the subject of elementary catastrophe theory. Owing to the condition (1.9), catastrophes of this type will be referred to as static. A catastrophe will be defined as a change in a set of critical points of a function V occurring on a continuous change of parameters $\mathbf{c}$. As will be shown later, the condition for occurrence of a catastrophe is expressed in terms of second derivatives of a function V, $\partial^2 V/\partial \psi_i \partial \psi_j$.

In the case when a change in control parameters is accompanied by a qualitative alteration of the form of solutions to equation (1.6) and condition (1.9) is not met, we deal with a dynamical catastrophe. It appears that in gradient systems (1.8) certain dynamical catastrophes important from a standpoint of applications, for example the so-called Hopf bifurcation, cannot occur. The description of such dynamical catastrophes thus requires utilization of equations for non-gradient autonomous systems — the systems of the type (1.6), where the condition (1.9) is not obeyed.

Let us now turn to a brief discussion of the examination of systems (1.6) by catastrophe theory techniques. To begin with, in systems (1.6) such static catastrophes may occur for which the condition (1.9) is satisfied. It follows from equations (1.6), (1.9) that in the stationary state of an autonomous system the condition

$$f_i(\psi; c) = 0, \quad i = 1, ..., N \tag{1.11}$$

analogous with the condition (1.10) for gradient systems, is fulfilled. Hence, the investigation of stationary states of equation (1.6) is reduced to a mathematical problem of examination of zero points of the mapping $\mathbf{f} = (f_1, ..., f_N)$. An investigation of the effect of changes of control parameters $\mathbf{c}$ on the properties of a set of solutions to equations (1.11) is the subject of singularity theory, which is a direct generalization of elementary catastrophe theory [cf. equations (1.11), (1.7) and (1.10)].

A typical example of a static catastrophe in the system (1.6) is the change in number of solutions of the system of equations (1.11) on varying $\mathbf{c}$ (a catastrophe of this type is called bifurcation). The condition of occurrence of a catastrophe of this type is expressed in terms of derivatives of functions f_i, $\partial f_i / \partial \psi_j$.

When the condition (1.9) is not met in (1.6), we deal with dynamical catastrophes. In some cases, for example for the so-called Hopf bifurcation, dynamical catastrophes may be examined by static methods of elementary catastrophe theory or singularity theory (Chapter 5). General dynamical catastrophes, taking place in autonomous systems, are dealt with by generalized catastrophe theory and bifurcation theory (having numerous common points). Some information on general dynamical catastrophes will be provided in Chapter 5.

1.3 EXPERIMENT

The notion of structural stability is also derived from the methodology of a scientific experiment. The possibility of verification of the experiment under slightly different conditions by another scientist presumes the resistance of a given experiment, its course and results, to perturbations of the initial data and conditions of its performance. One may thus agree with René Thom that "the hypothesis of structural stability is a postulate resulting from the very essence of any scientific observation".

As we shall demonstrate below, the necessity of description of phenomena of a discontinuous nature, proceeding with a continuous variation of control parameters, follows from the experimental studies. Thus, the problem of description of such processes ceases to be an academic problem.

We shall now describe a few phenomena from the area of physics, chemistry and biology, revealing features of a catastrophe, to illustrate some problems, both experimental and theoretical, of a new type, appearing in these areas. They may stimulate theoretical studies aiming at systematization, elucidation and prediction of phenomena of this type. It should be pointed out that an interest in such problems dates back to a relatively recent period, which was partly due to the lack of suitable theoretical tools. The following examples seem to have nothing in common:

(1) a soap film,
(2) the liquid-vapour system,
(3) a beam of radiation falling on a lens,
(4) a population of insects,
(5) the heartbeat,
(6) a chemical reaction,
(7) a non-linear recurrent equation.

However, in all these systems the phenomena of a catastrophic nature may be observed, as described below.

(1) Let the investigated system consist of a soap film stretched on two rings made from thin wire (Fig. 3).

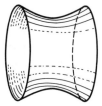

Fig. 3. Soap film stretched on wire rings.

A control parameter, varied by an experimenter, is the distance between parallelly placed rings. By the state of the system is meant the shape of the film. Let us assume that the initial distance between the rings is small (smaller than the ring radius R). The changes in the system taking place on increasing the distance between the rings l are shown in Fig. 4.

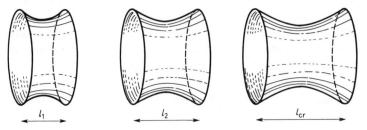

Fig. 4. Changes in a shape of the surface of a soap film.

On increasing l the film narrows, its shape remaining qualitatively the same all the time until the critical distance $l_{cr} \cong 1.3\,R$ is reached. Upon attaining the distance $l = l_{cr}$, the film begins to narrow spontaneously by itself (at a constant distance between the rings), bursts and jumps on side surfaces of the rings (Fig. 5).

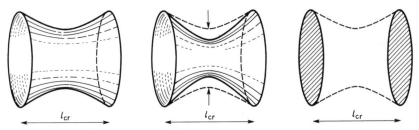

Fig. 5. Catastrophe of a soap film.

The described phenomenon is irreversible: a decrease in the distance l below l_{cr} does not apparently result in a change of the new state of the film. The following important property of the system should be emphasized: when the distance l is only slightly smaller than the critical distance, the system is very sensitive to disturbances. A very small increase of l (above the critical value) or a slight perturbation of the film (e.g. moving the system) are sufficient to initiate spontaneous change in the state of the film described above.

(2) The system consists of a liquid being in equilibrium with vapour at a set temperature T, pressure p and volume V. Control parameters are

pressure and temperature, while the state of the system is represented by density of the substance (proportional to $1/V$). At some temperature values the investigated fluid is gaseous (low density) for all pressure values, whereas at some other temperature values the liquid–vapour equilibrium is established. If at a certain temperature and pressure the system contains liquid (and some amount of vapour), then lowering of pressure may result in boiling of the liquid. If the pressure is decreased slowly, then the transition from the state of liquid–vapour equilibrium to the state of boiling occurs suddenly. Similarly, if the system contains only vapour, then the pressure increase at a constant temperature may result in spontaneous condensation of the vapour. The phenomena described above may be represented by means of phase diagrams (Fig. 6).

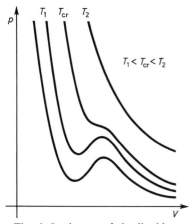

Fig. 6. Isotherms of the liquid–vapour system.

As follows from the diagram, at some pressure values (below the critical temperature T_{cr}) two states of the system are possible: (a) large volume (low density) corresponding to the vapour state; (b) small volume (high density) corresponding to the liquid state. In the vicinity of the point corresponding to the liquid–vapour transformation, the system exhibits sensitivity to perturbations and the more so the smaller is the difference between the system temperature and the critical temperature T_{cr}. In physical terms, the sensitivity of the system to perturbations in the vicinity of the critical point is manifested by critical opalescence — fluctuations of the fluid density cause that it scatters light (becomes turbid).

(3) In a case of a beam of rays passing through an optical system, spots of higher light intensity can be observed. In a geometrical optics approximation, these are spots of superposition of paths of two or more light rays — the so-called caustics. For example, light rays reflected from a cylindrical screen (this can be, for example, an internal surface of a vessel) form a caustic of the shape depicted in Fig. 7.

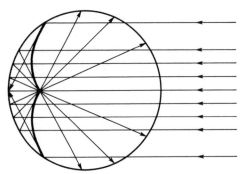

Fig. 7. Light rays reflected from a cylindrical screen.

Upon changing the position of the screen with respect to a light source (which corresponds to the change in control parameters), the observed image alters qualitatively. A catastrophe corresponds to a change in the form of a caustic appearing with a change in control parameters (Fig. 8).

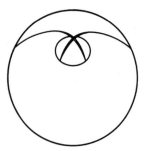

Fig. 8. Change in the form of the caustic on altering the screen position.

Another simple experiment may be performed with a converging lens (magnifying glass). Let a be the distance from an object to the lens, b — the distance from an image to the lens, and f be the focal length (Fig. 9).

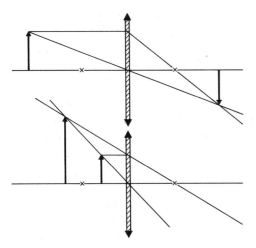

Fig. 9. Images of an object for a converging lens.

When $a > f$, a real inverted image is observed, whereas for $a < f$, the observed image is virtual and erect. The state of the system when $a = f$ (the case of an image at infinity) is sensitive to arbitrarily small disturbances — such a state of the system will be called a sensitive state. A catastrophe — the change in the form of an image — occurs at a continuous alteration of a control parameter, which is the distance from an object to the lens.

(4) The investigated system is a certain model population of insects,

(4.1) Insects lay eggs in autumn, from which individuals of a new generation are hatched in spring, none of the insects of the old generation surviving winter. There exists an optimum in the abundance of such a population. Too small a number of individuals of the new generation may bring about an extinction of the whole population after several years as a result of the activity of predators or losses due to an unfavourable influence of low temperatures on eggs in winter. In turn, an excessively large number of insects is also disadvantageous, since the overcrowding of insects results in a ready evolution of diseases, scarcity of food, increase of number of parasites and predators feeding on insects.

One of control parameters may be the ability of insects of a given population to reproduce (e.g. the number of eggs laid by an individual) or the average temperature in winter (determining the possibility of survival of undamaged eggs). A state variable is the abundance of population of a consecutive generation of adult individuals, e.g. in midsummer. It appears that in the existing populations of insects one may observe, depending on the values of control parameters, population extinction, stationary state, or oscillations in the population abundance having a period of a dozen or so years. A catastrophe will imply a change in the state of a population, for example the transition from a stationary state to an oscillatory state.

(4.2) Changes in the population abundance may be described theoretically. Let x_i be the abundance of i-th population of insects. One may anticipate that in the simplest case the following relationship between x_{i+1} and x_i will be satisfied:

$$x_{i+1} = f(x_i; c_1, c_2, ...) \tag{1.12}$$

where c_1, c_2 are control parameters, while f is a certain function (for example, determined experimentally).

(5) The investigated system is the heart. A state variable may be the length of the cardiac muscle: in the course of heartbeat the cardiac muscle contracts and decontracts, varying periodically its length. The heartbeat is controlled by arterial blood pressure (or by cardiac muscle tone directly related to arterial blood pressure) and biochemically, since the cardiac contraction is caused by the electrical impulse, generated in some biochemical reactions. Depending on the values of control parameters, several states of the heartbeat may be distinguished: normal heartbeat, ventricular or atrial fibrillation or flutter, etc. An overdose of a condiment, for example coffee, or a large loss of blood may change control parameters to such an extent that a catastrophe occurs, i.e. the transition from the normal heartbeat to the state of ventricular flutter or fibrillation.

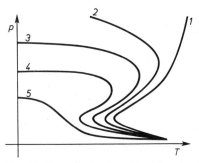

Fig. 10. Curves of self-ignition for the H_2/O_2 mixture resulting from electrical discharges: 1 — self-ignition region, $2-5$ — self-ignition regions for increasing intensity of electrical discharges.

(6) Examples of catastrophes which can be observed in chemical reactions will be discussed in more detail.

(6.1) The examined system consists of a mixture of gaseous hydrogen and oxygen. Under certain conditions, such a mixture may undergo a spontaneous ignition, whereas under some other conditions the self-ignition is not observed. The dependence of occurrence of self-ignition on pressure and temperature (which will be regarded as control parameters) for the H_2/O_2 mixture is illustrated by experimental curves (Fig. 10).

A variation in control parameters may be accompanied by a change in the state of the system (catastrophe): the system may pass from the state where self-ignition is impossible to the state in which self- ignition may take place.

(6.2) The system under study consists of the following reagents: malonic acid $CH_2(COOH)_2$, sodium bromate $NaBrO_3$ (commonly containing small amounts of bromide NaBr, or specially purified), cerium sulphate $Ce(SO_4)_2$, sulphuric acid and water (playing the role of solvent). In addition, the system may contain ferroin — a coloured redox indicator, greatly facilitating observation of changes occurring in the system. Behaviour of the system may be followed in a closed system (e.g. in a graduated cylinder) or in an open system (in a flow reactor); at the same time, the system may be homogenized by stirring. The system behaviour strongly depends on control parameters, these being (among other factors): the initial concentration of the reagents, stirring rate, temperature and, in the case of a flow reactor, retention time.

About 25 intermediates are formed in the reaction, while the final products are CO_2 and bromo derivatives of malonic acid. Ferroin, a redox indicator, contains the system Fe(II)/Fe(III). The oxidized form is blue, whereas the reduced form is red.

(6.2.1) The experiment carried out under homogeneous conditions, that is with stirring of the solution, will be described first. On mixing the reagents, a blue colour of the solution is observed. The solution, stirred throughout the experiment, gradually changes its colour, passing through shades of violet to red and remaining in this state for some time. A state parameter followed in the experiment is the relative concentration [Fe(II)/Fe(III)], determining a colour of the solution (in the case of absence of the indicator, a state parameter could be the concentration ratio [Ce(III)]/[Ce(IV)] or $[Br^-]$.

After some time has elapsed, the solution rapidly turns its colour to blue.

The described phenomenon repeats periodically with a period dependent on control parameters. The state of the system described above occurs in a rather broad range of control parameters. However, at an appropriate change of control parameters, oscillations of the solution colour may disappear. Such a change in the state of the system we will regard as a catastrophe.

(6.2.2) Another straightforward experiment is evolution of the same system without stirring; under such conditions the diffusion of reagents in the solution begins to play a decisive role. The experiment is carried out in the following manner. When the entire solution, prepared as previously, has a uniform colour, a small amount of the solution is transferred onto a Petri dish. In the system thus prepared, propagation of wavefronts, for example in the form of blue circles against the red background, is observed.

(6.2.3) Still other phenomena can be observed in a flow reactor. At properly selected concentrations of the components flowing into the reactor and at a suitable flow rate of the reagents through the reactor, a catastrophe is observed in the form of transition from the state of periodic oscillations of the solution colour to the state of random oscillations. For the parameter values close to critical values (for which transition to the state of chaotic changes in the solution colour takes place) the system exhibits sensitivity to perturbations of control parameters.

(6.2.4) General dynamical properties of the system described in Section 6.2.1 may be conveniently followed by monitoring the concentration of just one reagent, for example the Br^- ion (by an appropriate ion-selective electrode) or Ce^{3+} (colorimetrically). It is convenient to measure Br^- concentration at constant time intervals t. The measured concentrations will be equal to:

$$[Br^-](t_1), \ [Br^-](t_4 + t), \ [Br^-](t_1 + 2t), \dots$$

where t_1 is the time of the first measurement. Let us introduce designations $[Br^-](t_i) = x_i, t_i = t_1 + (i - 1)t$. The measurements of the quantity x_i reveal that the following recurrent equation

$$x_{i+1} = g(x_i; v) \tag{1.13}$$

is obeyed, where v is the flow rate of the reagents through the reactor (control parameter), whereas g is an experimentally determined function.

(7) Considerations presented in Sections 4.2 and 6.2.4 lead to a conclu-

sion that the properties of the following model [this is a simple variant of equations (1.12), (1.13)].

$$x_{i+1} = h(x_i) = 4bx_i(1 - x_i), \quad i = 1, 2, \dots \tag{1.14}$$

should be analysed, where b is a control parameter, $b \in [0, 1]$, and x_i is a state variable of the system, $x_i \in [0, 1]$. The iterative process described by the above equation reduces to computation of $x_2, x_3, \dots$ etc., on the basis of a known initial value of the variable $x = x_1$ and for a set value of the control parameter b. Computation of consecutive values of the sequence $\{x_i\}$ is conveniently performed by microcomputer. A character of the sequence $\{x_i\}$, that is the state of the system, strongly depends on the b value, while it is almost independent of the x_1 value. For example, when $0 \leqslant b \leqslant 1/4$ the sequence x_n approaches zero, when $1/4 < b \leqslant 3/4$ the sequence x_n approaches the value $1 - 1/4b$, whereas for b somewhat larger than $3/4$ the sequence x_n oscillates between certain two values (depending on b). On further increasing the parameter b, the oscillations become increasingly more complex. For most values of $b > 0.892486 \dots$ the sequence of x_n values is chaotic (nonperiodic).

1.4 SUMMARY

All the examples described above may be considered in uniform terms. In all the cases, control parameters may be readily given. These are varied by an experimenter or evolve with time, influencing the state of a system determined by other parameters — state variables of the system. In the described examples the behaviour of a specific system is qualitatively similar within a certain region of control parameters, whereas at some other values of these parameters (critical values) a distinct change in the system state takes place. In the vicinity of critical values of parameters the system exhibits a sensitive dependence on control parameters. Such phenomena result from the non-linearity of the system. These type of abrupt changes in the state of a system at a continuous variation of a control parameter will be referred to as catastrophes. It should be pointed out that the change in the state of a system in the case of a catastrophe always occurs with a finite rate and the process of "jumping" to a new state is not usually described by catastrophe theory. For example, for a distance between the rings increasing from $l = l_{cr} - \varepsilon$ (Fig. 5), where ε is an arbitrarily small positive number, to the value $l = l_{cr} + \varepsilon$, the system changes the state (at the point $l = l_{cr}$ the state of the system has a discontinuity and is a sensitive state). The change in shape of a soap film apparently occurs with a finite rate.

Commonness of the occurrence of such phenomena in non-linear systems — and the majority of interesting and non-trivial models are non-linear — has stimulated Reńe Thom to search for their universal description. The existence of a general theory of discontinuous change in the state of a system would have far-reaching consequences. The possibility of prediction of catastrophes for a specific model, or at least the classification of catastrophes which could occur, constitutes a strong incentive for the research in the area of catastrophe theory.

Examples 1–7 (subchapter 1.3), and particularly examples 6.2.1–6.2.4 concerning chemical reactions, will be examined in detail in further part of the book. However, even a cursory analysis of them leads to a conclusion that there are many analogies in behaviour of the described systems, for example an analogy of models 4, 6.2.4, 7, see equations (1.12), (1.13), (1.14). The analogy will become more evident upon comparing the form of functions f, g, h occurring in these equations. It turns out that the behaviour of a real population of insects may be modelled by setting $f(x) = 4ax(1 - x)$, where a is a control parametr. Similarly, experimentally determined dynamics of a chemical reaction (examples 6.2.1–6.2.4) may be well described by setting $g(x) = 4cx(1 - x)$, where c is a control parameter depending on the flow rate v of the reagents through a reactor. In other words, in equations (1.12), (1.13), (1.14) appears the same function (exact to within a proportionality constant). This conclusion accounts for the usefulness of investigation of the possibility of appearance of catastrophes in the abstract example 7.

The chemical systems described in example 6.2 are very interesting and at the same time important from the practical and theoretical standpoint. For example, there appeared a possibility of modelling periodical biological processes, controlled by the so-called biological clocks, by means of oscillating biochemical reactions. As will be shown later, in a model of the heartbeat (example 5) and in a model of an oscillating chemical reaction (example 6.2.3), catastrophes of an identical type may occur.

The similarity of Fig. 6 (model 2 of the phase transition in the liquid–vapour system) and 10 (self-ignition of the H_2/O_2 mixture, example 6.1) may suggest an analogy between catastrophes occurring in these systems; such an analogy actually takes place.

The examples describing catastrophes of a soap film may also find use in chemistry or biology. The shape of a soap film or membrane stretched on a given contour results from the condition of the minimal surface tension (since the surface tension is proportional to the area of the film or membrane). Hence, such a film has the shape of surface with a minimal surface area (for a given contour restricting it), that is a minimal surface.

From the above remarks follow applications of investigation of the states of minimal surfaces and possible changes in their shapes (catastrophes) on altering some control parameters. For example, foams may be formed at a phase boundary, thus determining the rate and character of reagent transport at the interface. Some small marine animals, such a *Radiolaria*, are built of skeletons covered with membranes. Thus, the structure of foam at the phase boundary or the shape of a radiolarian result from the same condition of a minimal surface area of the foam stretched at the interface or the membrane stretched on the skeleton of a radiolarian.

From the above considerations follows the significance of studies of possible states of minimal surfaces and likely changes in their shape (catastrophes) on varying some control parameters, similar to the catastrophes described in example 1 of Section 1.3 or more elaborate.

1.5 EVOLUTION OF CATASTROPHE THEORY AND ITS FOUNDERS

The program of catastrophe theory has been formulated by Thom. The fundamental theorems of elementary catastrophe theory have been proven by Thom, Mather and Arnol'd. A large contribution to this theory has been carried in by Zeeman, who also found many practical applications of the theory. Arnol'd and Berry have demonstrated the existence of a close relationship between elementary catastrophe theory and optics and found numerous uses for this relationship.

The most important notions of elementary catastrophe theory, such as structural stability, sensitive state, codimension, universal unfolding, have formed the basis for generalizations going beyond gradient systems (elementary catastrophe theory). A most significant share to generalized catastrophe theory has been contributed by Arnol'd, Stewart, Golubitsky, Schaeffer and Guckenheimer.

Catastrophe theory is not yet a closed area and it has been continuously evolved, finding new applications. It should be mentioned that a trend of unification of streams of catastrophe theory and other theories of non-linear phenomena, such as bifurcation theory, may be observed. If a theory including the entirety of non-linear phenomena ever evolves, then supposedly the theory of processes of a loss of a stability type (catastrophes) will owe much to the program of Thom's catastrophe theory.

Bibliographical Remarks

The program of catastrophe theory has been formulated in the Thom's book *Structural Stability and Morphogenesis*. The philosophical, mathematical and experimental incentives to the program of catastrophe theory may be found there. On the position of catastrophe theory among methods of investigation of stability of solutions of non-linear equations (models) one may learn from a book by Thompson. Stewart's and Zeeman's papers on the perspectives of the development of generalized catastrophe theory are also worth study.

A paper by Davis deals more extensively with the problems associated with computability. Technical aspects of the program of catastrophe theory are discussed in the Gilmore's book. A detailed bibliography on the experimental investigation of catastrophes is given in Chapters 3, 5 and 6. Hence, we shall now confine ourselves to calling the reader's attention to a paper by Swinney *et al.* on modelling the dynamics of a chemical reaction described in example 6.2.4 (Section 1.3) and to a May's paper concerning the dynamics of the non-linear model in example 7 and the behaviour of a model population of insects, example 4.

The references pertaining to chemical reactions will be given in Chapters 5 and 6. An exception, apart from a paper by Swinney, is the cited article by Winfree, describing the history of the Belousov–Zhabotinskii oscillatory reaction. The remaining references deal with philosophical problems.

References

M. Davis, "What is computation", in *Mathematics Today. Twelve Informal Essays*, L.A. Steen (Ed.), Springer-Verlag, New York–Heidelberg–Berlin, 1979.

F. Engels, *Dialektik der Natur*, Dietz Verlag, Berlin, 1952.

R. Gilmore, *Catastrophe Theory for Scientists and Engineers*, J. Wiley and Sons, New York, 1981.

W. Köhler, *Gestalt Psychology*, London, 1930.

R. M. May, "Simple mathematical models with very complicated dynamics", *Nature*, **261** (June 10), 459 (1976).

R. H. Simoyi, A. Wolf and H. L. Swinney, "One dimensional dynamics in a multi-component chemical reaction", *Phys. Rev. Lett.*, **49**, 245 (1982).

I. Stewart, "Beyond elementary catastrophe theory", *Mathematics and Computation*, **14**, 25 (1984).

R. Thom, *Structural Stability and Morphogenesis*, Benjamin-Addison-Wesly, New York, 1975.

J. M. T. Thompson, *Instabilities and Catastrophes in Science and Engineering*, J. Wiley and Sons, New York, 1982.

M. Wertheimer, *Productive Thinking*, New York–London, 1943.

A. T. Winfree, "The prehistory of the Belousov–Zhabotinskii oscillator", *J. Chem. Educ.*, **61**, 661 (1984).

E. C. Zeeman, "Bifurcation and catastrophe theory", *Contemporary Math.*, **9**, 207 (1982).

E. C. Zeeman, "Stability of dynamical systems", *Nonlinearity*, **1**, 115 (1988).

Chapter 2

Elementary Catastrophe Theory

2.1 INTRODUCTION

As explained in Section 1.2, the simplest field of applications of catastrophe theory are gradient systems (1.8). In the case of gradient systems, static catastrophes obeying the condition (1.9) can be studied by the methods of elementary catastrophe theory. Let us recall that a fundamental task of elementary catastrophe theory is the determination how properties of a set of critical points of potential function $V(\mathbf{x}; \mathbf{c})$ depend on control parameters $\mathbf{c}$. In other words, the problem involves an examination in what way properties of a set of critical points (denoted as M and called the catastrophe manifold)

$$M = \{(\mathbf{x}; \mathbf{c}): \ \nabla_x V(\mathbf{x}; \mathbf{c}) = 0\} \tag{2.1a}$$

$$\mathbf{x} = (x_1, ..., x_n), \ \mathbf{c} = (c_1, ..., c_k) \tag{2.1b}$$

depend on the control parameters $\mathbf{c}$, where ∇_x stands for the vector of partial derivatives with respect to the variables $\mathbf{x}$

$$\nabla_x = (\partial/\partial x_1, ..., \partial/\partial x_n) \tag{2.2}$$

For example, a set of critical points for a potential function of the form $(n = 1, \ k = 2)$

$$V = (x; a, b) = 1/4\,x^4 + 1/2\,ax^2 + bx \tag{2.3}$$

is given by the equation

$$M = \{(x, a, b): \ \mathrm{d}V/\mathrm{d}x = x^3 + ax + b = 0\} \tag{2.4}$$

hence, it is a certain surface in the three-dimensional space $(x, \ a, \ b)$.

A special attention in our considerations will be paid to the functions, further called structurally stable whose form, determined by the properties of a set of critical points, does not qualitatively change under not too large a perturbation. As explained in Sections 1.2, 1.3, the significance of a structurally stable function is related to the requirement of reproducibility of experiments. When we want to describe reproducible phenomena, the description of an experiment must contain structurally stable functions.

The fundamental tasks of elementary catastrophe theory may now be formulated:

one should find all classes of potential functions $V(x;c)$, containing: (1) equivalent functions (having the same form); (2) structurally stable functions.

Equivalent potential functions will be defined as the functions having identical sets of critical points. Such a definition of equivalence of potential functions implies that, as follows from the state equation (1.8) and equations (1.9), (1.10), potential functions of the same form describe physical systems having the same sets of stationary points.

The problems of structural stability and equivalence are associated with some technical problems which have to be resolved. These are the following problems:

(1) the problem of determinacy: at which point should the Taylor expansion of a given function be truncated so that the obtained series be a good approximation of this function near to the examined point;

(2) the problem of a (universal unfolding: what perturbations of a given function my change the nature of the investigated critical point of this function (that is, change local properties of the function near to the critical point);

(3) problem of classification: what are possible types of critical points of a function.

The relationship between problems 1–3 and the problems of structural stability and equivalence is a follows. To be able to divide potential functions into classes (containing equivalent functions), the definition of equivalence is required. Next, the knowledge of all essential perturbations of the function being studied (i.e. those which may change the nature of is critical point) is necessary to find whether it is structurally stable and to possibly extend it to a structurally stable function. If, for example, the function $V(x) = x^3$ perturbed by an addition of the function ax (where a is a constant) changes its local properties, then it may not be structurally stable. In such a case, the structural stability of the family of functions $V(x; a) = x^3 + ax$, containing (for $a = 0$) the original function, should be studied. As will be shown, there are no significant disturbances of such a potential function (i.e. those changing its character in the vicinity of critical points), that is the function $V(x; a)$ is structurally stable. The function $V(x; a) = x^3 + ax$ is called a universal unfolding of the function $V(x) = x^3$, whereas its term ax is referred to as a deformation.

The problem of determinacy results, on the one hand, from the trend of reducing the investigated functions and, on the other, from the fact that potential functions being fitted to experimental data are always approximate. Therefore, the information about the point of truncation of the Taylor series of a function having a critical point of presumed properties is essential.

2.2 POTENTIAL FUNCTIONS OF ONE VARIABLE

2.2.1 Introduction

A solution of the technical problem described above will be discussed first for the simplest case of a function of one state variable, the number of parameters being unlimited. Hence, we examine potential functions of the form $V(x; \mathbf{c})$, $\mathbf{c} = (c_1, ..., c_k)$, where x and c_i $(i = 1, ..., k)$ are real variables. To simplify notation we shall assume that the critical point is $x = 0$.

Local properties of a function $V(x)$ (in cases where it does not lead to misunderstanding we shall omit the dependence on $\mathbf{c}$) depend on its first and second derivative. Three basic cases may be distinguished:

(I) $V'(x) \neq 0$ for $x = 0$;

(II) $V'(x) = 0$ for $x = 0$, whereas $V''(x) \neq 0$ at $x = 0$;

(III) $V'(x) = 0$, $V''(x) = 0$ for $x = 0$.

Case I signifies that a function V does not have a critical point at $x = 0$. Case II corresponds to a so-called nondegenerate critical point, in case III the point $x = 0$ will be called a degenerate critical point. Further investigations of local properties of a function $V(x)$, aiming at the solution of problems 1–3 (determinacy, unfolding and classification), will account for the usefulness of distinguishing the basic cases I–III.

It will become evident later that catastrophes are associated with degenerate critical points of functions: only in this case may a change of differential type in a function (change in the set of its critical points – a catastrophe) take place on varying control parameters. We shall see that functions having points of type I or II are structurally stable, while functions having critical points of type III are structurally unstable. Hence, functions of type III cannot occur in the description of a physical process; on the other hand, the description of a catastrophe requires the use of just such functions. We shall demonstrate that the solution to this problem consists in embedding a function of type III in a family of functions, dependent on some parameters — a family of functions is commonly more stable structurally than a single function. The difference between the original function and a structurally stable family of functions will be called a universal unfolding of the original function.

2.2.2 Local properties of a function of one variable

To examine local properties of a function $V(x)$, let us expand it in a Taylor series in the neighbourhood of the point $x = 0$:

$$V(x) = V(0) + xV'(0) + 1/2x^2V''(0) + 1/3!x^3V'''(0) + ... \qquad (2.5)$$

The Taylor series for cases I–III distinguished above (Section 2.2.1) have the form:

$$\text{I} \quad V(x) = V(0) + xV'(0) + ... \qquad (2.6a)$$

$$\text{II} \quad V(x) = V(0) + 1/2x^2V''(0) + ... \qquad 2.6b)$$

$$\text{III} \quad V(x) = V(0) + 1/6x^3V'''(0) + ... \qquad (2.6c)$$

For a sufficiently small x (i.e. in a close vicinity of the point $x = 0$) the terms of higher order may be neglected in equations (2.6); at the same time, if $V'''(0) = 0$ then the first non-vanishing term of higher order should be left in (2.6c). In differential calculus it is proved that the thus truncated Taylor series is a good approximation of a function of one variable $V(x)$ near to the point $x = 0$.

Hence, the problem of determinacy 1 for a one-variable potential function has the following solution:

$$\text{I} \quad V(x) = xV'(0) \qquad (2.7a)$$

$$\text{II} \quad V(x) = 1/2x^2V''(0) \qquad (2.7b)$$

$$\text{III} \quad V(x) = 1/k!x^kV^{(k)}(0), \quad k \geqslant 3 \qquad (2.7c)$$

where the insignificant constant term $V(0)$ was disregarded in (2.7), while $V^{(k)}(0)$ is the first non-zero derivative in case III.

Equations (2.7) enable classification of local types of state functions in one variable (in previous considerations the dependence on parameters was not essential) and constitute the solution to problem 3 (classification). Case (2.7a) thus describes a regular (noncritical point), (2.7b) corresponds to the function having a maximum $(V''(0) < 0)$ or minimum $(V''(0) > 0)$, and the case (2.7c) corresponds to an inflection point for odd k values, whereas for even k values there is a maximum when $V^{(k)}(0) < 0$ or minimum when $V^{(k)}(0) > 0$.

TABLE 2.1

Classification of critical points of functions of one variable

x	I
$+x^2$	II A
$-x^2$	II B
$+x^4$	III A
$+x^6$	III A
$+x^8$	III A
$-x^4$	III B
$-x^6$	III B
$-x^8$	III B
x^3	III C
x^5	III C
x^7	III C

The simplest functions having critical points with properties I, II, III are those listed in Table 2.1, consecutive cases corresponding to a noncritical (regular) point I, nondegenerate minimum IIA, nondegenerate maximum IIB, degenerate minimum IIIA, degenerate maximum IIIB, inflection point IIIC (this is also a degenerate critical point). The functions having nondegenerate critical points IIA, IIB are sometimes called Morse functions.

The functions given in Table 2.1 are the simplest functions in one variable having at the point $x = 0$ local properties of a specific type. Sometimes these functions are called germs; more rigorously a germ (for example x^3) is defined as all functions locally equivalent (near to the point $x = 0$) to a given function (x^3).

All other functions of one variable, having at $x = 0$ the point of a given type, may differ from the functions given above only with respect to terms of higher order. Hence, Table 2.1 constitutes the ultimate solution of the problem of classification of a critical point type for functions of one variable.

EXAMPLE 2.1

$$V_1(x) = \cos(x), \qquad V_2(x) = \sin^3(x)$$

To examine local properties of these functions near to the point $x = 0$, consecutive derivatives of these functions at $x = 0$ are calculated. We obtain for the function $V_1(x)$: $V_1'(x) = -\sin(x)$, $V_1'(0) = 0$, $V_1''(x) = -\cos(x)$, $V_1''(0) = -1$, whereas for the function $V_2(x)$: $V_2'(x) = 3\sin^2(x)\cos(x)$, $V_2'(0) = 0$, $V_2''(x) = 6\sin(x)\cos^2(x) - 3\sin^2(x)$, $V_2''(0) = 0$, $V_2'''(x) = 6\cos^3(x) - 21\cos(x)\sin^2(x)$, $V_2'''(0) = 6$.

As follows from equations (2.7), the functions $W_1(x) = -1/2\,x^2$, $W_2(x) = x^3$ have locally, in the neighbourhood of the point $x = 0$, the same differential properties as the original functions $V_1(x)$, $V_2(x)$.

Indeed $W_1'(x) = -x$, $W_1'(0) = 0$, $W_1''(x) = -1$ and, moreover, $W_2'(x) = 3x^2$, $W_2'(0) = 0$, $W_2''(x) = 6x$, $W_2''(0) = 0$, $W_2'''(x) = 6$.

Finally, we conclude that at $x = 0$ the functions W_1, V_1 have a maximum and the functions W_2, V_2 have an inflection point.

2.2.3 Structural stability of critical points

The solution of the problems of determinacy and classification does not signify a complete characteristic of critical points. For example, from the intuitive standpoint cases (2.7c), corresponding to inflection points, $k = 3, 5, 7, \ldots$, are not equivalent. The simplest functions representing cases IIIA, IIIB, IIIC (see Table 2.1 above) will also turn out not to be structurally stable.

The problem of a more detailed characteristic of a critical point (including an examination of its structural stability) is related to the problem of unfolding (problem 2), i.e. to the determination of the effect of an arbitrary perturbation on a given critical point of the function being studied. Apparently, the simplest case is a critical point insensitive to perturbations — structurally stable IIA, IIB. In the case of structurally unstable degenerate critical points, the knowledge of changes in the local shape of a function in the neighbourhood of its critical point under perturbation is essential (the changes depend on the type of a degenerate critical point).

It will become evident that a solution of the problem of unfolding leads to a solution of the problem of structural stability; it allows, on the basis of knowledge of the effect of any perturbation on a given degenerate critical point IIIA–IIIC (see Table 2.1), to find a family of functions insensitive to perturbations and containing the simplest function with a critical point of the desired type.

EXAMPLE 2.2

$$V(x) = x^2$$

The function $V(x)$ has at $x = 0$ a nondegenerate critical point and, since $V''(0) = 2 > 0$, it is a minimum (case IIA). Let us examine the structural stability of the function $V(x)$ by adding to it any small perturbation

$$V_1(x) = V(x) + \varepsilon g(x), \quad 0 < \varepsilon \ll 1$$

(we shall assume that the perturbing function $g(x)$ is smooth near to the critical point; i.e. has continuous derivatives of arbitrarily high order).

The position of a critical point of the function $V_1(x)$, determined from the equation

$$V_1'(x) = 2x + \varepsilon g'(x) = 0$$

is close to the point $x = 0$ for a sufficiently small ε. Furthermore, the second derivative of the function V_1, $V_1''(x) = 2 + \varepsilon g''(x)$, is positive for a sufficiently small ε, that is the function $V_1(x)$ has a minimum nearby $x = 0$. Hence, the function $V(x)$ is structurally stable inasmuch as a small and practically any perturbation does not change the nature of is critical point (Fig. 11).

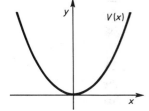
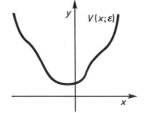

Fig. 11. Functions $V(x) = x^2$, $V(x; \varepsilon) = x^2 + \varepsilon g(x)$, $|\varepsilon| \ll 1$, $g(x) = x^2 + Ax + B$.

EXAMPLE 2.3

$$V(x) = x^3$$

At the point $x = 0$ the function $V(x)$ has a point of inflection — a degenerate critical point $(V'(0) = V''(0) = 0, \ V'''(0) = 0)$. Let us examine the structural stability of this function by perturbing it to

$$V_1(x) = V(x) + \varepsilon x^2, \quad 0 < \varepsilon \ll 1$$

The following relations hold for the function V_1:

$$V_1'(x) = 3x^2 + 2\varepsilon x, \qquad V_1'(0) = 0$$

$$V_1''(x) = 6x + 2\varepsilon, \qquad V_1''(0) = 2\varepsilon > 0$$

Thus, the perturbed function V_1 has at $x = 0$ a nondegenerate critical point — a minimum. Consequently, the function $V(x)$ is not structurally stable, since an addition of a small perturbation εx^2 modifies, for an arbitrarily small positive ε, its local properties near to the critical point (Fig. 12).

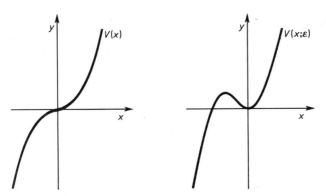

Fig. 12. Functions $V(x) = x^3$, $V(x; \varepsilon) = x^3 + \varepsilon x^2$, $0 < \varepsilon \ll 1$.

The above examples throw light on the problem of unfolding (problem 2) and the related problem of structural stability of a function of one variable. Although the function $V(x) = x^3$ is structurally unstable, one may hope that its modification having the form

$$V(x; a, b, ...) = V(x) + g(x; a, b, ...) \tag{2.8}$$

where $a, b, ...$ are parameters, containing for some values of these parameters the function $V(x)$ (i.e. for such values of the parameters at which the function g vanishes), will be structurally stable at a proper selection of the function $g(x; a, b, ...)$.

If a function g, called a deformation, contains all possible modifications of the critical point of a function $V(x)$, then any small perturbation of the function $V(x; a, b, ...) = V(x)\, g(x; a, b, ...)$ will not change the character of its critical point.

Such a structurally stable modification of the original function $V(x)$, $V(x; a, b, ...)$ will be referred to as a universal unfolding of the function $V(x)$.

The above considerations reveal the significance of the problem of unfolding: the knowledge of all essential perturbations of a function having a degenerate critical point of a given type is required.

We shall demonstrate how the problem of unfolding for a function of one variable can be resolved using the function $V(x) = x^3$ as an example. To begin with, using the already solved problem of determinacy the terms of order higher than three in the function g may be disregarded in (2.8)

$$V_1(x) = x^3 + Ax^2 + Bx + C \qquad (2.9)$$

Next, by setting $x \to x_1 = x + 1/3 \, A$, the quadratic term is removed from V_1:

$$V_1(x_1) = x_1^3 + (B - 1/3 \, A^2)x_1 + (C - 1/3 \, AB + 2/27 \, A^3) \qquad (2.10)$$

Neglecting in (2.10) the constant term (not changing the properties of the critical point) a function of the form

$$V(x; a) = x^3 + ax \qquad (2.11)$$

is obtained, where $a = B - 1/3 \, A^2$ and we have returned to the original form of the independent variable. Hence, a one-parameter family of functions of one variable $V(x; a)$ (2.11), describes all possible modifications of the critical point of the function $V(x) = x^3$ in the neighbourhood of its degenerate critical point (inflection point) $x = 0$.

It is now evident that the solution to the problem of unfolding (problem 2) would bring us nearer to the solution of the problem of structural stability. For each type of a degenerate critical point (7c) one should find

TABLE 2.2

Universal unfoldings of functions of one variable

Function	Deformation	Designation
x	—	—
$+x^2$	—	A_1
x^3	$+ax$	A_2
$+x^4$	$\pm(ax^2 + bx)$	A_3
x^5	$+ax^3 + bx^2 + cx$	A_4
$\pm x^6$	$\pm(ax^4 + bx^3 + cx^2 + dx)$	A_5
x^7	$+ax^5 + bx^4 + cx^3 + dx^2 + ex$	A_6

a universal unfolding of a suitable germ, i.e. the simplest function having a critical point of this type, see Table 2.1, and then check whether the obtained family of functions is structurally stable.

The problem of universal unfolding for the functions listed in Table 2.1 may be solved similarly to the problem of finding a form of the function $g(x; a, b)$ in (2.8). Universal unfoldings obtained in such a way are compiled in Table 2.2 (this is the first part of the Thom theorem, which pertains to potential functions in one state variable).

Structural stability of the above families of functions, containing structurally unstable functions with degenerate critical points at zero values of parameters, should now be examined. It will appear that embedding of structurally unstable functions (cases IIIA–IIIC, Table 2.1) in parametrized families of functions increased their structural stability: the functions given in Table 2.2 are structurally stable.

EXAMPLE 2.4

$$V(x; a) = x^3 + ax, \quad a < 0$$

One should examine whether the function

$$V(x; a, \varepsilon, g) = x^3 + ax + \varepsilon g(x)$$

has, for a small (positive) parameter ε and for any smooth function $g(x)$, the same local properties as the function $V(x)$. To this end, it is sufficient to consider the case of $g(x)$ of the form

$$g(x) = \varepsilon(x^2 + Ax + B), \quad 0 < \varepsilon \ll 1$$

since, as follows from the solution to the determinacy problem, perturbations containing higher powers of x cannot change the character of a critical point. Local properties of the function $V(x)$ and the function

$$V(x; a, \varepsilon, A, B) = x^3 + \varepsilon x^2 + (a + \varepsilon A)x + \varepsilon B$$

should thus be compared, where ε, $|\varepsilon A|$, $|\varepsilon B|$ are small compared with $|a|$, that is the solutions to the equations

$$V'(x; a) = 3x^2 + a = 0$$

$$V'(x; a, \varepsilon, A, B) = 3x^2 + 2\varepsilon x + (a + \varepsilon A) = 0$$

should be compared.

A simple inspection of solutions to the above equations leads to the

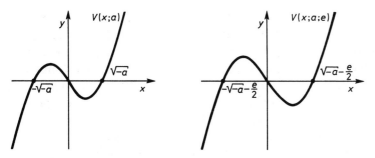

Fig. 13. Functions $V(x; a) = x^3$, $V(x; a, A, B, \varepsilon) = x^3 + \varepsilon x^2 + (a + \varepsilon A)x + \varepsilon B$, $|\varepsilon| \ll 1$.

conclusion that for a sufficiently small $|\varepsilon|$ the functions $V(x; a)$ and $V(x; a, \varepsilon, A, B)$ have critical points of the same type (see the plot Fig. 13). Hence, the function $V(x; a) = x^3 + ax$ is structurally stable.

The above considerations for a function of one variable are summarized in the Thom theorem (and more specifically, in its part dealing with functions of one variable) which classifies all structurally stable families of functions. The Thom theorem resolves the problems of determinacy, unfolding and classification. Table 2.2 lists structurally stable families of functions of one variable, containing (for zero values of parameters) the simplest functions with a critical point of a given type (see Table 2.1). Additional information on Thom functions of one variable is provided in Table 2.3.

TABLE 2.3

Properties of critical points of functions of one variable

Function	Designation	Point type	Number of parameters in universal unfolding
x	—	noncritical	0
$\pm x^2$	A_1	nondegenerate critical	0
x^3	A_2	degenerate critical	1
$\pm x^4$	A_3	degenerate critical	2
x^5	A_4	degenerate critical	3
$\pm x^6$	A_5	degenerate critical	4
x^7	A_6	degenerate critical	5

The designations given in Tables 2.2, 2.3 are frequently replaced with common names of potential functions of catastrophes and of the catastrophes themselves. In order $A_2, ..., A_6$ these are: fold, cusp, swallowtail, butterfly, wigwam. The potential functions A_1 are called Morse functions.

As follows from the Thom theorem, each structurally stable function of one state variable, dependent on at most five control parameters, must be equivalent to one of the functions listed in Table 2.2. Recall that functions are considered to be equivalent if they have identical sets of critical points. Another, equivalent definition, which will help to understand better the meaning of local equivalence of a function near to a critical point, is given below. Before that, however, let us examine two examples.

EXAMPLE 2.5

$$V_1(x) = x^2, \qquad V_2(x) = x^4 + x^2 + 1/4$$

Both functions have a minimum at $x = 0$: $V_1'(x) = 2x$, $V_1''(x) = 2$, $V_2'(x) = 4x^3 + 2x = 2x(2x^2 + 1)$, $V_2''(x) = 2 + 12x^2$. The shape of these functions in the vicinity of the critical point is shown in Fig.14.

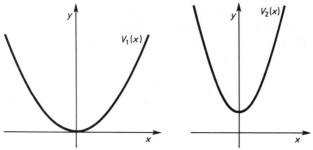

Fig. 14. Functions $V_1(x) = x^2$, $V_2(x) = x^4 + x^2 + 1/4$.

The plot of the function $V_2(x)$ near to the critical point $x = 0$ may be transformed into a plot of the function $V_1(x)$. This can be achieved by lowering the plot of $V_2(x)$ and changing a scale on the x-axis, i.e. by performing transformation of the form

$$V_2(x) = V_1[f(x)] + c \qquad (2.12)$$

where $f(x) = x(x^2 + 1)^{1/2}$, $c = 1/4$. The coordinate transformation $x \to \bar{x} = f(x)$ must be smooth (cannot produce additional critical points or remove them) and reversible. Thus, $f(x)$ has to have a non-zero derivative in the vicinity of a critical point. We check that $f'(x) = (1 + 2x^2)(1 + x^2)^{-1/2}$ does not vanish nearby $x = 0$.

The plot of the function $x \to f(x) = x(x^2 + 1)^{1/2}$ is shown in Fig. 15.

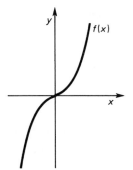

Fig. 15. Function $f(x) = x(x^2 + 1)^{1/2}$.

EXAMPLE 2.6

$$V_1(x) = x^2, \quad V_2(x) = x^6$$

Both functions have a minimum at $x = 0$, see Fig. 16.

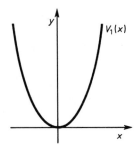 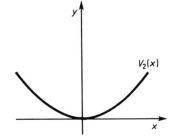

Fig. 16. Functions $V_1(x) = x^2$, $V_2(x) = x^6$.

The plot of $V_2(x)$ may be converted into the plot of $V_1(x)$ using the following change of variables $x \to \bar{x} = f(x) = x^3$:

$$V_2(x) = x^6 = V_1[f(x)] = (x^3)^2$$

However, the function $f(x) = x^3$, determining a change of variable, also has a critical point at $x = 0$, and such a change of variable is not allowed. Thus, the functions $V_1(x)$ and $V_2(x)$ are not locally equivalent in the neighbourhood of the point $x = 0$.

The above examples suggest the following definition of local equivalence of two functions $V_1(x)$ and $V_2(x)$. Two functions V_1, V_2 are locally equivalent in the vicinity of the point $x = x_0$ if there is such a constant c and such a smooth change of variable $x \rightarrow \bar{x} = f(x)$ that equation (2.12) is satisfied.

2.2.4 Catastrophes described by Thom potential functions of one variable

To conclude our considerations of functions of one variable, we shall examine the properties of several potential functions given in Table 2.2. In analysing the properties of these functions and the catastrophes described by them, the catastrophe manifold M defined by equation (2.1a) will be helpful. In the case of a function of one state variable it takes the form

$$M = \{(x; \mathbf{c}): dV/dx\,(x; \mathbf{c}) = 0\} \tag{2.13}$$

Evolution of the system on the surface of the potential minimum, i.e. on the catastrophe surface M, generated by a continuous change in the control parameter a, may now be investigated. A physical system tends towards minimization of potential energy, which is tantamount to the vanishing of the derivative $V_x'(x; a, b) = 0$. This condition implies that the system exhibits a tendency to remain on the surface of catastrophe M. A trajectory of the system on the catastrophe surface M may reflect a certain process, evolution of the system, generated by changes in control parameters and proceeding according to the principle of minimization of potential energy.

It is useful to distinguish in the catastrophe manifold M a subset Σ on which the function V has degenerate critical points since, as we shall see later, at these points a catastrophe takes place. The set Σ is thus given by the following equation:

$$\Sigma = \{(x, \mathbf{c}) \in M: V_x''(x; \mathbf{c}) = 0\} = \{(x; \mathbf{c}): V_x'(x; \mathbf{c}) = 0$$

$$V_x''(x; \mathbf{c}) = 0\} \tag{2.14}$$

A set containing points in the control parameters space, obtained by projection of points of the set Σ on the control parameters space, will be referred to as the bifurcation set B. Hence, the set B is given by the following equation:

$$B = \{(\mathbf{c}): \text{there exists } x \text{ such that } (x; \mathbf{c}) \in \Sigma\} \tag{2.15}$$

In addition, the catastrophe mapping Π — a projection of the point $(x; \mathbf{c})$ on the control parameters space

$$\Pi: \quad \Pi(x; \mathbf{c}) = (\mathbf{c}), \quad \text{or} \quad \Pi \Sigma = B \tag{2.16}$$

will prove useful.

We shall begin the investigation of Thom potential functions with the family of functions $V(x; a) = x^3 + ax$ corresponding to the fold catastrophe or A_2, see Table 2.2. The catastrophe manifold corresponding to this potential function is obtained from equation (2.13), $\mathbf{c} = (a)$,

$$M_2 = \{(x, a): 3x^2 + a = 0\} \tag{2.17}$$

where the subscript corresponds to designation of the catastrophe A_2. The set M_2 is depicted in Fig. 17.

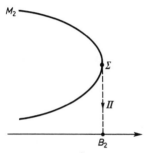

Fig. 17. Catastrophe surface M_2, singularity set Σ_2 and bifurcation set B_2 of the fold catastrophe (A_2).

According to equations (2.14), (2.15), the singularity set Σ_2 and the bifurcation set B_2 are of the form

$$\Sigma_2 = \{(x, a) \in M_2: x = 0\} = \{(0, 0)\} \tag{2.18}$$

$$B_2 = \{(a): \text{ there exists } x \text{ such that } (x, a) \in \Sigma_2\} = \{(0)\} \tag{2.19}$$

The sets Σ_2 and B_2 as well as the potential form in various regions of the bifurcation set are illustrated in Fig. 17.

Evolution of the system on the surface of the potential minimum, i.e. on the catastrophe surface M_2, generated by a continuous change of the control parameter a, may now be examined. A physical system described by the potential function $V(x; a, b)$ tends to stay on the catastrophe surface M_2 in agreement with the principle of minimization of potential energy.

An example of evolution of the system may look as that in Fig. 18.

The cusp catastrophe A_3 is described by the potential function (more exactly, the family of functions) $V(x; a, b) = 1/4\, x^4 + 1/2\, ax^2 + bx$ (coef-

Fig. 18. System evolution on the surface M_2.

ficients 1/4 and 1/2 were introduced to simplify the equations derived below), see Table 2.2. The catastrophe manifold M_3 is given by

$$M_3 = \{(x, a, b): x^3 + ax + b = 0\} \tag{2.20}$$

hence, it is a surface in the three-dimensional space (x, a, b), see Fig. 19.

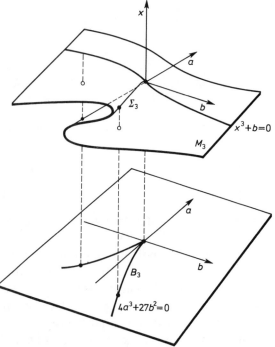

Fig. 19. Catastrophe surface M_3, singularity set Σ_3 and bifurcation set B_3 of the cusp catastrophe (A_3).

The set of degenerate critical points Σ_3 is expressed by

$$\Sigma_3 = \{(x, a, b): \; x^3 + ax + b = 0, \; 3x^2 + a = 0\} \tag{2.21}$$

Conditions defining the set Σ_3 are tantamount to the condition of having multiple roots by the polynomial $w(x) = x^3 + ax + b$, which follows immediately from the standard form: $w(x) = (x - x_1)(x - x_2)(x - x_3)$.

The set Σ_3 is shown in Fig. 19. On the surface M_3, at the point $(0, 0, 0)$ all three roots of the polynomial $w(x)$ are equal: $w_1 = w_2 = w_3 = 0$. The x variable may be eliminated from the system of equations defining the set Σ_3, thus obtaining the equation (this is the condition of existence of multiple roots for a third-order polynomial) $4a^3 + 27b^2 = 0$. Accordingly, the bifurcation set B_3 is given by

$$B_3 = \{(a, b): \; 4a^3 + 27b^2 = 0\} \tag{2.22}$$

The set B_3 is shown in Fig. 19. The figure also illustrates the form of potential $V(x; a, b)$ in various regions of the bifurcation set.

Let us now consider the behaviour of a physical system described by the potential function $V(x; a, b)$, whose evolution is generated by variations in the control parameters a, b. The system, in accordance with the general principle of minimization of potential energy, will tend to dwell on the catastrophe surface M_3. An exemplary trajectory of the system on the surface M_3 is shown in Fig. 20.

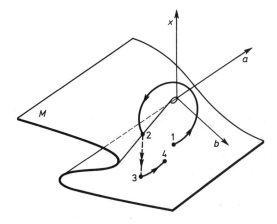

Fig. 20. System evolution on the surface M_3.

On a continuous change of control parameters a, b evolution of the system on the surface M_3 proceeds continuously from point 1 to point 2. However, on a further increase of the parameter b, the trajectory of the system must leave the surface of the potential energy minimum M_3. Hence, at point 2 a catastrophe — a qualitative change in the state of the system, takes place. The condition of the potential energy minimum requires the system to be present on the surface M_3. The system thus evolves possibly rapidly, from point 2 to point 3 along the path of shortest time. Elementary catastrophe theory does not describe the way of evolution along the path $2 \rightarrow 3$, but only to predicts the leaving of the surface M_3 for a time short compared with the time of evolution on M_3. Subsequently, a trajectory of the system proceeds on M_3 from point 3 to point 4. Note that the trajectory of the system on the surface M_3 has a discontinuity at point 2. The transition $2 \rightarrow 3$ has a locally irreversible character — in the vicinity of point 3 there is no return trajectory to point 2, whereas the transition is globally reversible along the path $3 \rightarrow 4 \rightarrow 1 \rightarrow 2$. The same process may also be shown in the control parameters space, see Fig. 21.

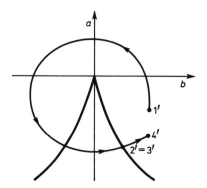

Fig. 21. System evolution in B_3.

The analysis of potential functions of one variable will finish with the examination of the swallowtail catastrophe A_4, to which corresponds the potential function $V(x; a, b, c) = 1/5\, x^5 + a/3\, x^3 + b/2\, x^2 + cx$ (again, we introduced for convenience of notation renormalization of coefficients in the catastrophe function, cf. Table 2.2).

The catastrophe manifold M_4 is of the following form:

$$M_4 = \{(x, a, b, c)\colon\ x^4 + ax^2 + bx + c = 0\} \qquad (2.23)$$

while the singularity set is given by

$$\Sigma_4 = \{(x, a, b, c): x^4 + ax^2 + bx + c = 0, 4x^3 + 2ax + b = 0\} \quad (2.24)$$

The manifold M_4 is a three-dimensional surface in a four-dimensional space and, consequently, it is difficult to describe. On the other hand, one may visualize the shape of the bifurcation set B_4 which is a two-dimensional surface embedded in a three-dimensional space.

To determine the form of the bifurcation set B_4, the x-variable should be eliminated from equations (2.24), defining the set Σ_4. This can be done in the following manner. The second of equations (2.24) is solved for x (a cubic equation has one or three real roots), which is then substituted to the first of equations (2.24). As a result, we obtain the equation

$$F(a, b, c) = 0 \qquad\qquad\qquad (2.25)$$

which is an equation of the surface defining the bifurcation set B_4. The shape of this set is illustrated in Fig. 22.

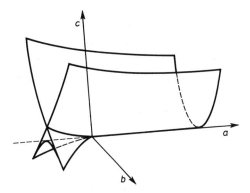

Fig. 22. Bifurcation set B_4 of the swallowtail catastrophe (A_4).

Noticeably, when the second of equations (2.24) has three real roots the set B_4 consists of three lobes, whereas when the equation has just one real solution then the set B_4 has only one lobe.

The next figure presents a section of the set B_4 with the plane $a = \text{const} < 0$ and the potential form in various regions of the section (Fig. 23).

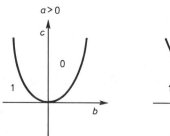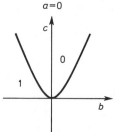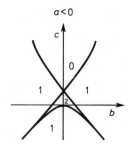

Fig. 23. Section of the bifurcation set B_4.

2.2.5 Summary

It follows from the examples examined above that in gradient systems catastrophes may occur only in a case when the system is described by a potential function having a degenerate critical point, for in this case the set Σ delimitating in M the functions of a various differential type is not an empty set. On exceeding by the system the set Σ on the catastrophe surface M, the change in a local type of a potential function $V(x; c)$, i.e. a catastrophe, takes place at a continuous change of control parameters.

The sets M, Σ, B, defined for families of functions of one variable at the beginning of Section 2.2.4, will play an equally vital role in the analysis of catastrophes occurring in systems described by potential functions dependent on two state variables. For the case of two state variables, the sets will be defined in Section 2.3.6.

2.3 FUNCTIONS OF TWO VARIABLES

2.3.1 Introduction

We shall now proceed to investigate potential functions $V(\mathbf{x}; \mathbf{c})$ dependent on many state variables and on parameters, where $\mathbf{x} = (x_1, ..., x_n)$, $\mathbf{c} = (c_1, ..., c_k)$.

Similarly to the case of functions of on variable, three fundamental situations may take place:

(I) at the investigated point, for example $\mathbf{x} = (0, ..., 0) \equiv 0$, the function does not have a critical point (such a point will be called regular point);

(II) the point $\mathbf{x} = 0$ is a structurally stable critical point (such a point will be called nondegenerate critical point);

(III) the point $\mathbf{x} = 0$ is a structurally unstable critical point (degenerate critical point);

furthermore, it will appear that properties I, II, III may be formulated in terms of (partial) derivatives of first and second order, analogously with the case of a function of one variable.

From the standpoint of elementary catastrophe theory, the functions having degenerate critical points are most interesting. As follows from Section 2.2, in gradient systems catastrophes may happen only in a case when the system is described by a potential function having a degenerate critical point.

Such a function may be included into a structurally stable parameter--dependent family of functions which will be considered to be a potential function. The state of a physical system will be determined from the condition of the minimum of a potential function having a degenerate critical point, defining the catastrophe surface M.

It will appear that on evolution of a system generated by continuous variations of control parameters and proceeding on the catastrophe surface a continuous evolution of the system on this surface is not always possible. In other words, in some circumstances a catastrophe must occur — the system has to leave for some time the surface of the potential minimum.

The critical point $\mathbf{x} = 0$ of a function of many variables $V(x_1, ..., x_n)$ may turn out to be a degenerate critical point solely due to dependence on some variables only. In such cases, separation of the function into two terms is possible

$$V(x_1, ... x_n) = V_{NM}(x_1, ..., x_m) + V_M(x_{m+1}, ..., x_n) \qquad (2.26)$$

(subscripts M and NM refer to the Morse and non-Morse term, respectively). The function V_{NM} has a degenerate critical point at the point $\mathbf{x}_{NM} = (x_1, ..., x_m) = 0$, while the point $\mathbf{x}_M = (x_{m+1} ...., x_n) = 0$ is a nondegenerate critical point of the function V_M. In such a case the investigation may be restricted to the function V_{NM} only, as only degenerate critical points may give rise to a catastrophe (see discussion at the end of Section 2.2). The method of separation of a function V into components V_{NM}, V_M will be discussed in Section 2.3.4.

In many cases interesting from the viewpoint of applicability the function V_{NM} will be a function of one or two variables. Accordingly, our considerations will be limited to potential functions V_{NM} dependent on two variables.

In a further part of our considerations we shall define a regular point, a nondegenerate critical point and a degenerate critical point. As promised earlier, we shall show how a potential function V is split into the term V_{NM} having degenerate critical points (interesting from the standpoint of elementary catastrophe theory) and the term V_M (not leading to the occurrence of catastrophes). The problems of determinacy 1, unfolding 2 and classification 3 (see Section 2.1) will be discussed next. The case of a function dependent on two variables will appear to be appreciably more complex than that of a function of one state variable. Consequently, we shall confine ourselves to the derivation of only a part of the results contained in the fragment of Thom's theorem pertaining to functions dependent on two state variables (see Appendix). It should be emphasized at this point what is achieved by solving in turn problems 1 and 2. In a case of a function having a degenerate critical point, a solution to the problem of determinacy 1 allows to find such an approximation of the examined function with a Taylor series that an additional perturbation will not alter the degenerate critical point into another degenerate critical point. On the other hand, a solution of the problem of universal unfolding for a function for which problem 1 has already been solved permits to find a structurally stable family of functions containing the investigated function.

The outline of the next part of the chapter is as follows. We shall discuss properties of a function in the vicinity of a regular point and a nondegenerate critical point. Then, we shall discuss the case of functions having degenerate critical points. It will turn out that a critical point may be degenerate with respect to only a part of its state variables and we shall demonstrate how a function having such a critical point can be separated into two terms, V_{NM} and V_M, in accordance with equation (2.26). Only the function V_{NM} has a degenerate critical point; this means a substantial simplification of the problem, for it is necessary for a catastrophe to occur that the potential have degenerate critical points (see Section 2.3.1).

Subsequently, examples of functions of two variables having degenerate critical points will be examined and the difficulties related to the problems of determinacy and unfolding discussed. We shall give a list of structurally stable families of functions of two variables, having degenerate critical points for some values of parameters on which they depend (this is the second part of the Thom theorem). Finally, we shall examine properties of potential functions in terms of possible catastrophes. For this purpose, the notions of catastrophe manifold M, singularity set Σ and bifurcation set B, introduced in Section 2.2, will be employed.

The chapter ends with an Appendix wherein certain very important concepts associated with properties of potential functions describing catastrophes will be given and, to some extent, the Thom theorem substantiated.

We shall now proceed to obtaining for functions of two variables the results analogous with those listed in Table 1 for functions of one variable. In other words, we shall find possible types of a local shape of functions in the neighbourhood of a certain point and the simplest functions having at the point $x = 0$ a specific local type. The accomplishment of this program corresponds to a solution of the problem of classification of possible types of a local shape of functions.

2.3.2 Regular points of potential functions

A regular point of a function of n variables $V(x_1, ..., x_n)$ is defined as the point not being a critical point of this function. In other words, at a regular point, let it be for example the point $\mathbf{x} = 0$, the function gradient (cf. definition (2.2)) does not vanish:

$$\nabla_x V(0, ..., 0) \neq 0 \qquad (2.27)$$

In the case of a function of one variable, $V(x)$, it could be represented in the vicinity of a regular point in the form $V(x) \cong x$, see equation (2.7a). In the case of a regular critical point of several variables, the function $V(x_1, ..., x_n)$ can be simplified in a similar way, without changing its local character nearby this point.

The definition of local equivalence of a function (2.12) suggests that a simpler form of a potential function with the same local properties may be sought by way of change of variables.

In that case, let us perform the following change of variables $\mathbf{x} \to \mathbf{x}'$ in a potential function satisfying the condition (2.27)

$$\begin{aligned}
x_1' &= V(x_1, ..., x_n) \\
x_2' &= a_{21}x_1 + a_{22}x_2 + ... + a_{2n}x_n \\
&\cdots\cdots\cdots\cdots\cdots\cdots\cdots\cdots\cdots\cdots \\
x_n' &= a_{n1}x_1 + a_{n2}x_2 + ... + a_{nn}x_n
\end{aligned} \qquad (2.28)$$

where a_{ij} are some, yet undefined, numerical coefficients.

The transformation $\mathbf{x} \to x'$ is an allowed change of variables when it does not alter the character of the investigated regular point $\mathbf{x}$ 0. This is the case when the Jacobian of transformation (2.28) does not vanish nearby the point $\mathbf{x} = 0$

$$\det \begin{vmatrix} \partial V/\partial x_1 & \partial V/\partial x_2 & \cdots & \partial V/\partial x_n \\ a_{21} & a_{22} & \cdots & a_{2n} \\ a_{31} & a_{32} & \cdots & a_{3n} \\ \multicolumn{4}{c}{\dotfill} \\ a_{n1} & a_{n2} & \cdots & a_{nn} \end{vmatrix} \neq 0 \qquad (2.29)$$

This condition can always be met by a suitable selection of the coefficients a_{ij} if the first row of the determinant does not vanish, that is when the function V has non-zero partial derivatives in the neighbourhood of the point $\mathbf{x} = 0$. In the case of a function of one variable we would have the transformation $x \to x' = V(x)$. To be an allowed change of variables, this has to be a one-to-one, i.e. invertible, transformation; the dependence $x = f(x')$ may then be determined which is possible when the function V has a non-zero first derivative nearby $x = 0$.

To recapitulate in the vicinity of a regular point a function of several variables may be unambiguously presented in the form $V(x') = x'$ on performing an appropriate change of variables, equation (2.28). For a function of one variable, this case I, Table 1.

EXAMPLE 2.7

$$V(\mathbf{x}) = x_1 x_2{}^2 + x_1$$

At the point $\mathbf{x} = (0, 0)$, partial derivatives of the function V are

$$\partial V/\partial x_1 \, (0, 0) = \{x_2{}^2 + 1\}_{(0,0)} = 1$$
$$\partial V/\partial x_2 \, (0, 0) = \{2x_1 x_2\}_{(0,0)} = 0$$

and hence the point $\mathbf{x} = (0, 0)$ is not a critical point of the function V. Transformation (2.28) has the form

$$x_1 \to x_1' = x_1 + x_1 x_2{}^2$$
$$x_2 \to x_2' = a_{21} x_1 + a_{22} x_2$$

The Jacobian matrix of this transformation at the point $(0, 0)$ has the form

$$\begin{vmatrix} 1 & 0 \\ a_{21} & a_{22} \end{vmatrix}$$

and its determinant is equal to a_{22}. The condition for vanishing of the Jacobian will be satisfied, for example, on selecting $a_{21} = 0$, $a_{22} = 1$. Finally, the sought transformation is given by

$$x_1 \to x_1' = x_1 + x_1 x_2^2$$
$$x_2 \to x_2' = x_2$$

and the potential function in new coordinates is $V(\mathbf{x}') = x_1'$.

We shall now proceed to the case of a function of several variables having a structurally stable critical point. For a function in one variable this was the case of a Morse function — II, Table 1. We shall also give a condition sufficient for the stability of a critical point which can be expressed in terms of derivatives $\partial^2 V / \partial x_1 \partial x_j$.

2.3.3 Morse lemma

When the function $V(\mathbf{x})$ has a critical point at $\mathbf{x} = 0$, the transformation defined above, (2.28), does not determine a one-to-one change of variables nearby the point $\mathbf{x} = 0$, because the first row of the Jacobian matrix vanishes and, consequently, so does its determinant. As a result, such a function cannot be represented as $V(\mathbf{x}') = x_1'$ in the vicinity of its critical point. In a case, however, when the critical point $x = 0$ is structurally stable (a precise criterion will be provided in further part of this section), the function V may be reduced in the vicinity of this point to a simpler form.

The method of reduction of the form of the function $V(x)$ in the case of a structurally stable critical point will be discussed for a function V dependent on two variables. The presented results are a special case of the Morse lemma.

Let the function $V(\mathbf{x})$, $\mathbf{x} = (x, y)$, having at $\mathbf{x} = (0, 0)$ a critical point, have the following general form:

$$V(x, y) = ax^2 + bxy + cy^2 + [\text{higher-order terms}] \tag{2.30}$$

Indeed, at $\mathbf{x} = 0$ a function V of the form (2.30) has a critical point, since at this point vanish first-order partial derivatives:

$$\partial V/\partial x\,(0,0) = \{2ax + by + [\text{higher-order terms}]\}_{(0,0)} = 0$$

$$\partial V/\partial y\,(0,0) = \{6x + 2cy + [\text{higher-order terms}]\}_{(0,0)} = 0 \qquad (2.31)$$

We shall assume about the function V that in the neighbourhood of the point $\mathbf{x} = 0$ the Hessian matrix V_{ij} has a non-zero determinant

$$\det[V_{ij}(0)] \equiv \det \begin{vmatrix} \partial^2 V/\partial x^2 & \partial^2 V/\partial x\,\partial y \\ \partial^2 V/\partial x\,\partial y & \partial^2 V/\partial y^2 \end{vmatrix}_{(0,0)} = \begin{vmatrix} 2a & b \\ b & 2c \end{vmatrix} = 4ac - b^2 \neq 0$$

$$(2.32)$$

A more exact meaning of this condition will soon become clear. We shall only state that the condition signifies structural stability of a critical point. The point $\mathbf{x} = 0$, on meeting conditions (2.31), (2.32), will be called a nondegenerate critical point. The above definitions are also applicable to the case of a function dependent on any number of variables (the matrix V_{ij} is the matrix of second derivatives of the function V).

To simplify the form of the function V in the vicinity of its critical point, let us make the following change of variables:

$$x \to \bar{x} = Ax + By$$

$$y \to \bar{y} = Cx + Dy \qquad (2.33)$$

assuming about transformation (2.33) (this is the diagonalization of the quadratic form $ax^2 + bxy + cy^2$ from equation (2.30) that its determinant does not vanish: $AD - BC \neq 0$.

On the basis of equations (2.30), (2.33) we obtain

$$V(\bar{x}, \bar{y}) = \lambda_1 \bar{x}^2 + \lambda_2 \bar{y}^2 + [d\bar{x}^3 + e\bar{x}^2\bar{y} + f\bar{x}\bar{y}^2 + g\bar{y}^3] + \dots$$

$$\lambda_{1,2} = (a + c) \pm [(a - c)^2 + b^2]^{1/2} \qquad (2.34)$$

When the transformation matrix (2.33) is orthogonal $(A^2 + B^2 = 1, C^2 + D^2 = 1, AC + BD = 0)$, λ_1, λ_2 are called eigenvalues of the quadratic form $ax^2 + 2bxy + cy^2$ or of the matrix $V_{ij}(0)$. Since the matrix V_{ij} is symmetric, the eigenvalues are real.

A new transformation of variables may now be performed:

$$\bar{x} \to x' = \bar{x} + (A_{20}\bar{x}^2 + A_{11}\bar{x}\bar{y} + A_{02}\bar{y}^2) + \dots$$

$$\bar{y} \to y' = \bar{y} + (B_{20}\bar{x}^2 + B_{11}\bar{x}\bar{y} + B_{02}\bar{y}^2) + \dots \qquad (2.35)$$

selecting the coefficients A_{ij}, B_{ij} in such a way that in new variables the function V has the form

$$V(x', y') = \lambda_1 x'^2 + \lambda_2 y'^2 \tag{2.36a}$$

Let us note that the function (2.36a) contains only the Morse term (see equation (2.26)). The variables x', y' may be additionally renormalized:

$\lambda_1 > 0, \ \lambda_2 > 0$

$$\tilde{x} = (\lambda_1)^{1/2} x', \qquad \tilde{y} = (\lambda_2)^{1/2} y', \qquad V(\tilde{x}, \tilde{y}) = \tilde{x}^2 + \tilde{y}^2$$

$\lambda_1 > 0, \ \lambda_2 < 0$

$$\tilde{x} = (\lambda_1)^{1/2} x', \qquad \tilde{y} = (-\lambda_2)^{1/2} y', \quad V(\tilde{x}, \tilde{y}) = \tilde{x}^2 - \tilde{y}^2$$

$\lambda_1 < 0, \ \lambda_2 > 0$

$$\tilde{x} = (-\lambda_1)^{1/2} x', \quad \tilde{y} = (\lambda_2)^{1/2} y', \qquad V(\tilde{x}, \tilde{y}) = -\tilde{x}^2 + \tilde{y}^2$$

$\lambda_1 < 0, \ \lambda_2 < 0$

$$\tilde{x} = (-\lambda_1)^{1/2} x', \quad \tilde{y} = (-\lambda_2)^{1/2} y', \quad V(\tilde{x}, \tilde{y}) = -\tilde{x}^2 - \tilde{y}^2 \tag{2.36b}$$

It appears that the condition (2.32) always allows to find transformation coefficients (2.35) and to reduce the function V in variables $\tilde{x}$, $\tilde{y}$ to one of the functions (2.36b). By analogy with case II (see Table 1) for a function of one variable, such a form of the function is called the Morse form.

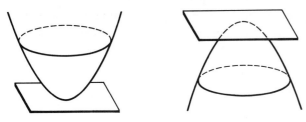

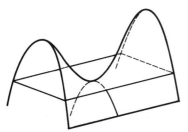

Fig. 24. Surfaces $x^2 + y^2$ (minimum), $-x^2 - y^2$ (maximum), $x^2 - y^2$ (saddle point).

The function $x^2 + y^2$ corresponds to a minimum, the function $-(x^2 + y^2)$ describes a maximum, the functions $x^2 - y^2$ or $-x^2 + y^2$ represent a saddle point, see Fig. 24.

Recapitulating, in the vicinity of a nondegenerate critical point a function in two (and, generally, in several) variables is locally equivalent to the quadratic form (2.36). The condition of lack of degeneracy of a critical point, $4ac - b^2 \neq 0$, implies non-vanishing of both the eigenvalues λ_1, λ_2 of the matrix of second derivatives (Hessian) computed at the point (0, 0). Note also that the condition (2.32), $\det[V_{ij}(0)] = 0$, signifies vanishing of a least one eigenvalue. This is a sufficient condition for degeneracy of the critical point $\mathbf{x} = 0$.

To illustrate the above theorems, consider the following example.

EXAMPLE 2.8

$$V(x, y) = x^2 + xy + 1/4\,y^3 - 1/16\,y^4$$

The function V has a critical point at $x = 0$. The Hessian matrix V_{ij} is of the form

$$V_{ij} = \begin{vmatrix} 2 & 1 \\ 1 & 3/2\,y - 3/4\,y^2 \end{vmatrix}$$

and at the point $x = 0$, $y = 0$ has a non-zero determinant; hence, it is a nondegenerate critical point. The eigenvalues λ_1, λ_2 are calculated from the equation

$$\det[V_{ij}(0) - \lambda] = \begin{vmatrix} 2 - \lambda & 1 \\ 1 & -\lambda \end{vmatrix} = -\lambda(2 - \lambda) - 1 = 0$$

$\lambda_1 = 1 + \sqrt{2} > 0$, $\lambda_2 = 1 - \sqrt{2} < 0$.

Note that the function V may be written in the form

$$V(x, y) = (x + 1/2\,y)^2 - (1/2\,y)^2 + 1/4\,y^3 - 1/16\,y^4$$

hence, transformation (2.33) (note that this is not an orthogonal transformation)

$$x \to \bar{x} = x + 1/2\,y$$
$$y \to \bar{y} = 1/2\,y$$

reduces the function V in new variables to the form (2.34)

$$V(\bar{x}, \bar{y}) = \bar{x}^2 - \bar{y}^2 + 2\bar{y}^3 - \bar{y}^4$$

Since the function $V(\bar{x}, \bar{y})$ can be written in the equivalent form

$$V(\bar{x}, \bar{y}) = \bar{x}^2 - (\bar{y} - \bar{y}^2)^2$$

the transformation (2.35) may thus be expressed as

$$\bar{x} \to x' = \bar{x}$$
$$\bar{y} \to y' = \bar{y} - \bar{y}^2$$

and the function V in variables x', y' is of the form (2.36b)

$$V(x', y') = x'^2 - y'^2$$

At the end of our considerations on nondegenerate critical points it should be emphasized that functions of the form (2.36) have a structurally stable critical point. Accordingly, in the systems described by Morse potential functions, catastrophes cannot occur. The structural stability of the function (2.36) will be demonstrated in the Appendix.

2.3.4 Splitting lemma

When the matrix V_{ij} has a zero determinant at a critical point, then this is a degenerate point. The degeneracy may turn out, however, to occur only due to a part of variables of the function $V(\mathbf{x})$. This is revealed by vanishing of only some of the eigenvalues of the matrix of second derivatives (Hessian) at the critical point $\mathbf{x} = 0$. For a function dependent on two variables, given by (2.30), upon transformation (2.33) we obtain

$$V(\bar{x}, \bar{y}) = \lambda_1 \bar{x}^2 + 0\bar{y}^2 + (a\bar{x}^3 + b\bar{x}^2\bar{y} + c\bar{x}\bar{y}^2 + d\bar{y}^3) + \dots \qquad (2.37)$$

As the coefficient at $\bar{y}^2$ vanishes, transformation (2.35) cannot be performed, for the four unknown coefficients a, b, c, d in equation (2.37) cannot be determined by means of the three coefficients A_{20}, A_{11}, A_{02}.

On the other hand, the function V may be split into two parts, V_M and V_{NM}, see equation (2.26), only the function V_{NM} having a degenerate critical point. Transformation (2.35) applied to the function (2.37) reduces it to a simpler form (but not so simple as (2.36))

$$V(x', y') = \lambda_1 x'^2 + 0y'^2 + V_{NM}(y') \qquad (2.38)$$

Such a reduction is feasible due to a sufficient number of equations resulting from non-vanishing of λ_1.

Splitting a function into the components V_{NM} and V_M will be discussed using as an example the function V dependent on two variables.

EXAMPLE 2.9

$$V(x, y) = x^2 + xy + 1/4\,x^2 + 2xy^2$$

The matrix V_{ij} at the critical point $x = 0$ is given by

$$V_{ij} = \begin{vmatrix} 2 & 1 + 4y \\ 1 + 4y & 1/2 + 4x \end{vmatrix}_{(0,0)} = \begin{vmatrix} 2 & 1 \\ 1 & 1/2 \end{vmatrix}$$

Since $\det[V_{ij}(0)] = 0$, we deal with a degenerate critical point. The eigenvalues of the matrix V_{ij} are given by the equation

$$\det \begin{vmatrix} 2 - \lambda & 1 \\ 1 & 1/2 - \lambda \end{vmatrix} = \lambda^2 - 5/2\,\lambda = 0$$

and are thus equal to 0, 5/2.

Non-orthogonal transformation (2.33)

$$x \rightarrow \bar{x} = x + 1/2$$
$$y \rightarrow \bar{y} = y$$

reduces the function V to the following form:

$$V(\bar{x}, \bar{y}) = \bar{x}^2 + 0\bar{y}^2 + 2\bar{x}\bar{y}^2 - \bar{y}^3$$

The next transformation, of type (2.35),

$$\bar{x} \rightarrow x' = \bar{x} + \bar{y}^2$$
$$\bar{y} \rightarrow y' = \bar{y}$$

yields

$$V(x', \bar{y}) = \{x'^2 + 0y'^2\} + \{-y'^3 - y'^4\} \equiv V_M(x') + V_{NM}(y')$$

We may now formulate the splitting lemma. When a function $V(x_1, ..., x_m, x_{m+1}, ..., x_n)$ has at $x = 0$ a degenerate critical point and not all the eigenvalues of the second derivative matrix V_{ij} are equal to zero: $\lambda_1 = ... = \lambda_m = 0$, $\lambda_{m+1}\lambda_{m+2} ... \lambda_n \neq 0$, then the function V can be split into the two components $V_{NM}(x_1, ..., x_m)$ and $V_M(x_{m+1}, ..., x_n)$. Furthermore, the term having a nondegenerate critical point, V_M, may be represented in suitable coordinates in the Morse form

$$V(x'_{m+1}, ..., x'_n) = \sum_{i=m+1}^{n} \lambda_i x_i'^2 \qquad (2.39)$$

As explained earlier, only the term V_{NM} may lead to the occurrence of a catastrophe in elementary catastrophe theory.

When all the eigenvalues of the Hessian matrix V_{ij} vanish, the splitting cannot be carried out.

2.3.5 Functions having degenerate critical points

We shall begin our considerations by exemplifying that in the case of functions having degenerate critical points the problems of determinacy and unfolding are considerably complicated compared with a function of one variable. Functions of two variables having a degenerate critical point, for example $x^2 y$, $x^2 + y^4$, $x^2 + y^5$, $x^5 + y^5$, have a number of surprising properties, differing considerably from functions of one variable (having a degenerate critical point) and still more, from the Morse functions in one or two variables. Examples illustrating the difficulties encountered when solving the problem of determinacy or the problem of unfolding will be given below.

If we add terms of higher order to a function of two variables having a degenerate critical point then, in contrast with the case of a function in one variable, the local character of the perturbed function in the neighbourhood of the degenerate critical point may be drastically changed.

EXAMPLE 2.10

$$V_1(x, y) = x^2 y, \qquad V_2(x, y; \varepsilon) = x^2 y + \varepsilon y^4, \qquad V_3(x, y; \varepsilon) = x^2 y + \varepsilon y^5$$
$$(0 < \varepsilon \ll 1)$$

The above functions have, at the point $(x, y) = (0, 0)$, a degenerate critical point, and for small ε the functions V_2, V_3 may be considered as a perturbation of the function V_1. However, the local shape of the functions V_1, V_2, V_3 near to the point $(0, 0)$ is, for an arbitrarily small ε, quite different (Fig. 25).

The zeros of the functions V_1, V_2, V_3, given by the equations: $x = 0$, $y = 0$; $y = 0$, $x^2 + \varepsilon y^3 = 0$, $y = 0$ are marked in the figure.

It follows from the above example that in the case of a function having a degenerate critical point, the terms of higher order cannot be arbitrarily

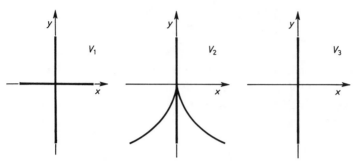

Fig. 25. Zeros of functions $V_1(x) = x^2 y$, $V_2 = x^2 y + \varepsilon y^4$, $V_3(x) = x^2 y + \varepsilon y^5$, $(\varepsilon > 0)$.

neglected, since an arbitrarily small perturbation of the function causes a degenerate critical point (points) to turn into a different type of a degenerate critical point (points). In such a case the function is structurally unstable. However, it will soon turn out that taking into account some higher-order terms allows us to neglect successive terms of higher order. For instance, the function $V(x, y) = x^2 y + y^4$, perturbed by the addition of higher-order terms, preserves its local character — see Appendix.

Similarly, finding a universal unfolding for a function having a degenerate critical point is not a trivial task. Recall that the problem of unfolding consists in finding all possible modifications of a degenerate critical point resulting from the added perturbation. At this point, we are considering the case of a function for which the problem of determinacy has already been solved. In other words, the added perturbation cannot change the nature (kind) of a degenerate critical point; it is possible, however, to convert a degenerate critical point into a nondegenerate point. For example, in the case of a function of one variable it sufficed to examine the effect of addition of terms of order lower than the order of a polynomial determining the type of a given critical point (see Table 1).

EXAMPLE 2.11
$$V(x, y) = x^5 + y^5$$

The above function has a degenerate critical point at (0, 0). Let us consider the following perturbation of this function

$$V(x, y) \rightarrow V(x, y; \varepsilon_{ij}) = x^5 + y^5 + \sum_{i, j} \varepsilon_{ij} x^i y^j$$

where $|\varepsilon_{ij}| \ll 1$.

It turns out that to describe all possible local modifications of the function $V(x, y)$ (in the vicinity of the point $(0, 0)$) caused by addition of a perturbation, the term $\varepsilon_{33}x^3y^3$ has to be included, among other terms, in the function $V(x, y; \varepsilon_{ij})$ (see example A7 in Appendix). This is surprising, as this term is higher by one order than that of the unperturbed function (in the case of a function of one variable the terms of order higher than the of a polynomial defining the type of a degenerate critical point, see Table 1, could be disregarded).

A conclusion may be drawn from the above example that when solving the problem of determinacy and the problem of unfolding for a function of two variables one should not be influenced by the results obtained for a function in one variable.

As mentioned above, we shall not describe at this point the method of solving the problem of determinacy for functions of two variables (see Appendix, A2). We shall confine ourselves to providing a list of the simplest potential function, having at $\mathbf{x} = (0, 0)$ a regular point, a degenerate critical point, for which the problem of determinacy has been solved. In other words, addition of a perturbation to the functions listed in Table 2.4 must not convert a degenerate critical point into another degenerate critical point. Table 2.4 (functions of two variables) is a counterpart to Table 2.1 (functions of one variable).

TABLE 2.4

Classification of critical points of two variables

x	$-$
$\pm(x^2 + y^2)$	A_1
$x^2 - y^2$	A_1
$x^2y - y^3\{x^3 - 3x^2y\}$	D_4^-
$x^2y + y^3\{x^3 + y^3\}$	D_4^+
$\pm(x^2y + y^4)$	D_5
$x^2y - y^5$	D_6^-
$x^2y + y^5$	D_6^+
$x^3 - y^4$	E_6^-
$x^3 + y^4$	E_6^+

The simplest functions compiled in Table 2.4, having degenerate critical points of a given type $(D_4^-, ..., E_6)$, are structurally unstable. As with to the case of functions of one variable, structurally stable functions having a critical point of a given type can be obtained by adding an appropriate function from Table 2.4 to a family of functions. In other words, the problem of universal unfolding has to be solved for functions $D_4^-, ..., E_6$. We shall now provide a list, see Table 2.5, of structurally stable families of functions dependent on two state functions and on a number of parameters (not more than five). The list of catastrophes depending on three and four parameters compiled by Thom has been extended by Arnol'd to include the case of five parameters. Partial substantiation of the form of the families of functions contained in Table 2.5 will be given in Appendix. Table 2.5 corresponds to Tables 2.2, 2.3 for functions of one variable.

TABLE 2.5

Universal unfoldings of two variables

Function	Deformation	No. of parameters	Designation
x	—	0	—
$\pm(x^2 \pm y^2)$	—	0	A_1
$x^2y - y^3$	$+ax^2 + by + cx$	3	D_4^-
$x^3 - 3xy^2$	$+a(x^2 + y^2) + bx + cy$	3	D_4^-
$x^3 + y^3$	$+axy + bx + cy$	3	D_4^+
$x^2y + y^3$	$+a(-x^2 + y^2) + bx + cy$	3	D_4^+
$\pm(x^2y + y^4)$	$\pm(ax^2 + by^2 + cx + dy)$	4	D_5
$x^2y - y^4$	$+ay^3 + by^2 + cx^2 + dx + ey$	5	D_6^-
$x^2y + y^5$	$+ay^3 + by^2 + cx^2 + dx + ey$	5	D_6^+
$x^3 - y^4$	$\pm(axy^2 + by^2 + cxy + dx + ey)$	5	E_6^-
$x^3 + y^4$	$\pm(axy^2 + by^2 + cxy + dx + ey)$	5	E_6^+

To a potential function in one variable (see Table 2.2) may be added, without a change in properties of the elementary catastrophe being described, the term M and to a potential function in two variables the term N:

$$M = \pm x_2^2 \pm x_3^2 \pm ... \pm x_n^2$$
$$N = \pm x_3^2 \pm x_4^2 \pm ... \pm x_n^2$$

i.e. any combination of Morse functions of the remaining state variables. Table 2.5 lists two, completely equivalent, frequently used parametrizations for the functions D_4^- and D_4^+. One may pass from the second to the first parametrization using a linear change of variables:

$$x \to -3^{-1/2}y$$

$$y \to x \ \text{for} \ D_4^{-}$$

$$x \to x + y$$

$$y \to -x + y \ \text{for} \ D_4^{+}$$

Elementary catastrophes of two variables also have common names. In order D_4^{-}, D_4^{+}, ..., E_6 these are: elliptic umbilic, hyperbolic umbilic, parabolic umbilic, second elliptic umbilic, second hyperbolic umbilic and symbolic umbilic catastrophes.

The Thom theorem (with the Arnol'd extension) resolves the problem of classification of families of structurally stable functions of one and two state variables, dependent no not more than five control parameters. Each of the families of functions occurring in the Thom theorem (see Tables 2.2, 2.5) contains a function having a degenerate critical point (see Tables 2.1, 2.4). Families of functions are parametrized in such a way that the function having a degenerate critical point is obtained for zero values of the parameters. The number of parameters occurring in the unfolding of a given function is called the codimension; it is an important notion of catastrophe theory whose meaning will be given in the Appendix.

Let us recall that the Thom functions are structurally stable due to accounting for appropriate lower-order terms (the problem of unfolding); as a result, the insensitivity to perturbations containing terms of low order is achieved, and, due to a proper selection of the form of functions having degenerate critical points (the problem of determinacy, see Tables 2.1, 2.4), the insensitivity to perturbations containing higher-order terms is attained.

From the Thom theorem (with Arnol'd modification) follows the very important conclusion that each function of one or two variables having a critical point, dependent on not more than five parameters and structurally stable must be equivalent, in the sense of the definition given above (see equation (2.12)), to one of the functions given in the list of functions of elementary catastrophes (Tables 2.2, 2.5).

2.3.6 Elementary catastrophes for functions of two variables

We shall now proceed to the examination of catastrophes modelled by Thom potential functions of two state variables. This will be done using the notions of the catastrophe manifold M, singularity set Σ and bifurcation set

B, employed in Section 2.2 to investigate potential functions of one variable. In the case of functions of two variables the catastrophe manifold M is given by the equation (cf. equation (2.1a))

$$M = \{(x, y, c_1, ..., c_k): \partial V/\partial x = \partial V/\partial y = 0\}. \tag{2.40}$$

The set Σ is such a subset of the set M that the function V has degenerate critical points at points belonging to Σ. The sufficient condition for a critical point to be degenerate is vanishing of the Hessian matrix determinant, i.e. vanishing of at least one eigenvalue of the Hessian matrix (the catastrophe functions in two variables have both the eigenvalues equal to zero).

In other words, the set Σ is given by

$$\Sigma = \{(x, y, c_1, ..., c_k) \in M: \det|\partial^2 V/\partial x \partial y| = 0\} \tag{2.41}$$

Points in the control parameters space, for which the function $V(x, y, c_1, ..., c_k)$ has a degenerate critical point, belong to the bifurcation set B. Hence, the set B has the form:

$$B = \{(c_1, ..., c_k): \text{ there exist such values of } x \text{ and } y \text{ that}$$
$$(x, y, c_1, ..., c_k) \in \Sigma\} \tag{2.42}$$

Using the definition of catastrophe mapping (projection) Π (cf. (2.16))

$$\Pi(\mathbf{x}, \mathbf{c}) = (\mathbf{c}) \tag{2.43}$$

equation (2.42) can be written as

$$B = \Pi\Sigma \tag{2.44}$$

The description of catastrophes corresponding to functions of two variables will be begin with a hyperbolic umbilic catastrophe D_4^+, which is represented by the potential function

$$V(x, y; a, b, c) = x^3 + y^3 + axy + bx + cy$$

The catastrophe manifold M, according to equation (2.40), is of the form

$$M = \left\{(x, y, a, b, c): \begin{array}{l} 3x^2 + ay + b = 0 \\ 3y^2 + ax + c = 0 \end{array}\right\} \tag{2.45}$$

The set Σ is expressed by equation (see (2.41))

$$\Sigma = \left\{(x, y, a, b, c): \begin{array}{l} 3x^2 + ay + b = 0, \quad 3y^2 + ax + c = 0 \\ 36xy - a^2 = 0 \end{array}\right\} \tag{2.46}$$

since

$$\det \begin{vmatrix} \partial^2 V/\partial x^2 & \partial^2 V/\partial x \partial y \\ \partial^2 V/\partial x \partial y & \partial^2 V/\partial y^2 \end{vmatrix} = \det \begin{vmatrix} 6x & a \\ a & 6y \end{vmatrix} = 36xy - a^2 \qquad (2.47)$$

The name of the catastrophe derives from the last equation in (2.46), which is the equation of a hyperbola.

The set M may be parametrized by expressing the parameters b, c by means of equations (2.45)

$$M = \{(x, y, a, b, c) = (x, y, a, -ay - 3x^2, -ax - 3y^2)\} \qquad (2.48)$$

On parametrizing the equation $36xy - a^2 = 0$ in the following manner (for non-zero a)

$$a = 6\alpha, \qquad x = \alpha\xi, \qquad y = \alpha/\xi \qquad (2.49)$$

parametrization of the set Σ is obtained

$$\Sigma = \{(x, y, a, b, c) = (\alpha\xi, \alpha/\xi, 6\alpha, -6\alpha^2\xi^2, -6\alpha^2\xi - 3\alpha^2/\xi^2)\} \qquad (2.50)$$

Parametrization of the bifurcation set B follows from (2.50) as a projection on the parameter space:

$$B = \{(a, b, c) = (6\alpha, \alpha^2(-6/\xi - 3\xi^2), \alpha^2(-6\xi - 3/\xi^2))\} \qquad (2.51)$$

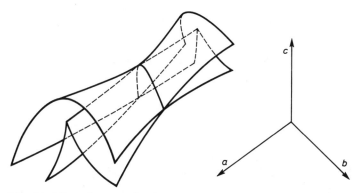

Fig. 26. Bifurcation set of a hyperbolic umbilic catastrophe $(D_4{}^+)$.

The bifurcation set is shown in Fig. 26. An elliptic umbilic catastrophe is described by the potential function

$$V(x, y; a, b, c) = x^3 - 3xy^2 + a(x^2 + y^2) + bx + cy$$

The catastrophe manifold M, according to (2.40), is of the form

$$M = \left\{(x, y, a, b, c): \begin{array}{l} 3x^2 - 3y^2 + 2ax + b = 0 \\ -6xy + 2ay + c = 0 \end{array}\right\} \tag{2.52}$$

The set M may be parametrized by expressing the parameters b, c by means of equations (2.52)

$$M = \{(x, y, a, b, c) = (x, y, a, -2ax + 3y^2 - 3x^2, -2ay + 6xy)\} \tag{2.53}$$

The set Σ is determined from condition (2.41)

$$\Sigma = \left\{(x, y, a, b, c): \begin{array}{l} 3x^2 - 3y^2 + 2ax + b = 0, \ -6xy + 2ay + c = 0 \\ (3x)^2 + (3y)^2 = a^2 \end{array}\right\} \tag{2.54}$$

since

$$\det \begin{vmatrix} \partial^2 V/\partial x^2 & \partial^2 V/\partial x \partial y \\ \partial^2 V/\partial y \partial x & \partial^2 V/\partial y^2 \end{vmatrix} = \det \begin{vmatrix} 6x + 2a & -6y \\ -6y & -6x + 2a \end{vmatrix} = 4a^2 - 36(x^2 + y^2)$$

$$= 4\{a^2 - [(3x)^2 + (3y)^2]\} \tag{2.55}$$

Transforming variables x, y to polar coordinates

$$x = (a/3) \sin(v)$$
$$x^2 + y^2 = (a/3)^2 \tag{2.56}$$
$$y = (a/3) \cos(v)$$

we obtain on the basis of (2.56) and parametrization (2.53):

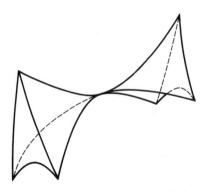

Fig. 27. Bifurcation set of an elliptic umbilic catastrophe (D_4^-).

$$\Sigma = \{(x, y, a, b, c)$$
$$= (x, y, a, 1/3\,a^2[\cos(2v) - 2\sin(v)], 1/3\,a^2[\sin(2v) - 2\cos(v)])\}$$

$$(2.57)$$

Parametrization of the bifurcation set B is obtained from (2.57) by projection of Σ on the parametr subspace:

$$B = \{(a\,b, c) = (a, a^2/3[\cos(2v) - 2\sin(v)], a/3[\sin(2v) - 2\cos(v)])\}$$

new parameters being a, v. The set B is shown in Fig. 27.

2.4 FINAL REMARKS

At the end of a brief review of elementary catastrophes, in which we discussed Morse functions in one and two state variables (A_1), catastrophes of one state variable A_2, A_3, A_4, and catastrophes of two state variables $D_4{}^+$, $D_4{}^-$, it should be noted that elementary catastrophes form a certain hierarchy according to the diagram shown below

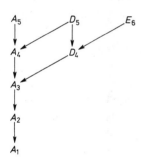

to the effect that the neighbourhood of a degenerate critical point of a catastrophe of lower order is contained within the neighbourhood of a degenerate critical point of a catastrophe of higher order.

The catastrophes described by Thom potential functions in two variables may be described in the control parameters space. The catastrophe takes place when a trajectory of the system on the catastrophe surface M, projected on the control parameters space, intersects the bifurcation set B (i.e. the projection of the set Σ).

Appendix

A2.1 INTRODUCTION

In the formulation of the Thom theorem occur such important concepts of catastrophe theory as: equivalence, determinacy, universal unfolding, codimension. Due to the vital role of these notions in catastrophe theory, we shall try to describe them in more detail. Let us add that the material presented in this Appendix is derived, to a large extent, from the papers of Mather.

Recall that a function is finitely determined, more specifically k-determined, when it may be locally replaced (that is approximated near to a given point) by the finite, k-term Taylor expansion. All functions of elementary catastrophes given by Thom and Arnol'd, see Tables 2.2, 2.5, are k-determined. This is a very important property of a function, allowing us to locally examine its characteristics (and, moreover, this is the necessary condition for a structural stability of the respective family of functions). As demonstrated in Example 2.10, the function $x^2 y$ is not k-determined. Hence, local investigation, in the vicinity of the point (0, 0), of a function whose first non-zero term of the Taylor expansion is $x^2 y$ without the knowledge of next terms of the Taylor expansion, is impossible. On the other hand, it follows from the Thom theorem that, for example, the function $V(x) = x^2 y + ay^4 + ...$ is k-determined and subsequent terms of the Taylor series may be neglected.

The notion of equivalence, see equation (2.12), permits to reduce the problem, in the case of a structurally stable potential function dependent on not more than five control parameters, to one of the cases occurring in the compilation of elementary catastrophes by Thom and Arnol'd (Tables 2.2, 2.5). Owing to the Thom theorem we can be sure that each k-parameter, for $k \leqslant 5$, structurally stable potential function can be reduced to some of the elementary catastrophe functions.

An essential concept appearing in the Thom theorem is codimension. Formally, it is the number of parameters present in a universal unfolding of a specific elementary catastrophe. The codimension has, however, a considerably greater significance. We shall introduce below the elements of a theory enabling the better understanding of the meaning of this notion and the calculation of the codimension of functions having degenerate critical points as well as finding the forms of a universal unfolding.

A2.2 ELEMENTS OF CODIMENSION THEORY

Let us assume that $V: R^N + R$ is a smooth function defined near $\mathbf{x} = (x_1, ..., x_N) = 0$ (R stands for the set of real numbers — the real axis, while R^N denotes the set of vectors consisting of N real numbers — the real N-dimensional space).

E_N will stand for the set of all such functions. We shall define an expansion of the function $V \in E_N$ in a Taylor series as the jet jV:

$$jV = V(0) + \sum_i x_i (\partial V / \partial x_i)(0) + \sum_{ij} 1/2 x_i x_j (\partial^2 V / \partial x_i \partial x_j)(0) + ... \qquad (A1)$$

the above definition being understood quite formally; in other words, the series (A1) does not have to be convergent.

Let F_n be the set of all formal power series. If so, there is a map

$$E_N \rightarrow F_N$$

defined by the Taylor equation (A1), and assigning to functions from E_N the series from F_N. The k-jet $j^k V$ is the Taylor series up to and including terms of order k.

For example, for the function $\cos(x)$ its Taylor series and the four-order can be written in the form

$$j[\cos(x)] = 1 - 1/2 x^2 + 1/24 x^4 - 1/720 x^6 + ...$$

$$j^4[\cos(x)] = 1 - 1/2 x^2 + 1/24 x^4$$

The function V is said to be k-determined if every function U meeting the condition $j^k U = j^k V$ (i.e. having an identical Taylor expansion up to and including terms of order k) is equivalent, in the sense of definition (2.12), to the function V. In other words, a local character of the function V in the neighbourhood of $\mathbf{x} = 0$ is fully determined by k first terms of the Taylor series (modification of higher-order terms of the Taylor series, i.e. re-placement of V with U, does not change a local shape of V).

Thus, a k-determined function is equivalent to some polynomial in variables $x_1, ..., x_N$ of order k. For instance, the function $V(x) = \sin(x)$ in the sufficiently close vicinity of $x = 0$ may be approximated with an arbitrary accuracy by the function $U(x) = x$; in other words $\sin(x) \cong j^1[\sin(x)] = x$ for $x \cong 0$. We shall now define within E_N a subset m_N consisting of all functions $V \in E_N$ such that $V(0) = 0$.

For $|\mathbf{x}| \cong 0$, the absolute values of functions $V \in m_N$ are small. The subset m_N has the following properties:

If $V \in m_N$, and $U \in E_N$, then

$$VU \subset m_N, \text{ or } m_N E_N = m_N \qquad \text{(A2)}$$

because the functions from E_N cannot take infinite values at $\mathbf{x} = 0$, whereas the functions from m_N vanish at $\mathbf{x} = 0$; hence, the product VU vanishes at $\mathbf{x} = 0$.

Property (A2), $m_N E_N = m_N$, is a definition of an ideal: a set absorbing with respect to some property, here: with respect to the property of vanishing at $\mathbf{x} = 0$.

Consequently, the set m_N consists of such functions $V \in E_N$ that $V(0) = 0$. In that case $m_N{}^2$, or the set consisting of the products UV, U, $V \in m_N$, contains all functions belonging to E_N such that all first-order terms of their Taylor series vanish at $x = 0$. In other words, at $\mathbf{x} = 0$ vanish the differentials of these functions:

$$V \in m_N{}^2, \ dV(0) = \sum_i dx_i (\partial V / \partial x_i)(0) = 0$$

As a result, a smooth function vanishing at $\mathbf{x} = 0$ and having there a critical point belongs to $m_N{}^2$.

Similarly, $m_N{}^k$ contains such functions that for $V \in m_N{}^k$ the differentials $dV(0) = d^2 V(0) = \ldots = d^{k-1} V(0) = 0$. Hence, the functions belonging to $m_N{}^3$ have at $\mathbf{x} = 0$ a degenerate critical point.

EXAMPLE A.1

$$V(x) = \sin(x), \qquad U(x) = x$$

The functions V, $U \in m_N$, since $V(0) = 0 = U(0)$; similarly, the functions x^2, x^3, etc. also belong to m_N. To the set $m_N{}^2$ belong, among other functions, all functions $W = UV$, where U, $V \in m_N$. Thus, $x^2 \in m_N{}^2$, while x, $\sin(x) \in m_N{}^2$.

Apparently, $x \in m_N{}^2$, since it can be represented as a product of two functions vanishing at $x = 0$; $x = x^{1/2} \times x^{1/2}$. However, the function $U(x) = x^{1/2}$ does not belong to E_N, for it is not smooth at zero: $U'(x) = {}^{1/2} x^{-1/2}$ is not well defined at $x = 0$.

The ideals $m_N{}^k$ form an inclusive chain

$$\ldots \subset m_N{}^3 \subset m_N{}^2 \subset m_N \subset E_N$$

In the set of formal power seris F_N we shall distinguish the subset M_N of power series with a zero constant term $(V(0) = 0$ in equation (2.44)) − the set M_N is an analogue of the set m_N. Note that $M_N = j(m_N)$.

Similarly, the equality

$$M_N{}^k = j(m_N{}^k)$$

holds, that is $M_N{}^k$ is the set of formal power series not containing in the series (2.44) the first $k - 1$ terms ($M_N{}^2$ denotes the set of products of series, each of which belong to M_N, etc.).

In the case when a smooth function V, vanishing at zero, has a critical point at $\mathbf{x} = 0$, the set of functions of the form:

$$U_1(\partial V/\partial x_1) + \ldots + U_N(\partial V/\partial x_N) \equiv \triangle(V) \tag{A3}$$

where $U_1, \ldots, U_N$ are arbitrary functions belonging to E_N is an ideal, called the Jacobian ideal. Since $U_i \in E_N$, $V \in m_N$, hence $V(0) = 0$, $dV(0) = 0$ and, consequently, $\triangle(V) \subseteq m_N$.

EXAMPLE A.2

$$V(x) = x^k$$

For the function $V\colon R^N \to R$ we shall calculate the Jacobian ideal. According to definition (A3)

$$\Delta(V) = \text{set of al functions of the form } U(x)kx^{k-1}$$

where $U(x) \in E_1$.

Finally

$$\Delta(V) = M_1{}^{k-1}$$

In the Jacobian ideal of the function $V(x) = x^k$ there are no functions which would be proportional to monomials $x, x^2, \ldots, x^{k-2}$ (or whose first non-zero terms of the Taylor expansion would contain these monomials).

We may now define the codimension which measures (the complexity of) a singularity of a critical point of a function. Let us recall that the function $V(x)$ has a critical point at $\mathbf{x} = 0$, if $[\partial V/\partial x_i](0) = 0$ for $i = 1, \ldots, N$. Then, as we already know, $\Delta(V) \subseteq m_N$, where $\Delta(V)$ is the Jacobian ideal of V.

A codimension of a function V, $\mathrm{cod}(V)$, is defined as the number of monomials missing from $\Delta(V)$:

$$\mathrm{cod}(V) = \text{dimension of subspace complementing } \Delta(V) \text{ to } M_N \tag{A4}$$

The following theorem is valid. If either of $\mathrm{cod}(V)$ or $\mathrm{cod}(jV)$ is finite, then so is the other, and they are then equal. The theorem permits us to calculate the codimension of a function using its formal Taylor series.

EXAMPLE A.3

$$V(x) = x^k$$

As follows from Example A.2:

$$\Delta(V) = \{\text{function of the form } U(x)x^{k-1}; \ U(x) \in E_1\}$$

that is, $\Delta(V)$ contains all monomials x^{k-1}, x^k, x^{k+1}, ...

The space M_1 contains polynomials consisting of the following monomials:

$$x, x^2, ..., x^{k-1}, x^k, x^{k+1}, ...$$

Hence, the space complementing $\Delta(V)$ to M_1 contains monomials x, $x^2, ..., x^{k-1}$. The dimension of this space, and thus the codimension of the function V, is $k-2$.

Apparently, in the neighbourhood of the critical point of the function $V(x) = x^k$, i.e when $dV/dx(x) = kx^{k-1} \cong 0$, we can be sure that the functions whose first non-zero terms of the Taylor expansion begin with the term proportional to x^{k-1} (or still higher order) are small in this vicinity, whereas we have no certainty that the functions of order x^l, $1 < k-1$, are small. In other words, the perturbed function $V_1(x) = x^k + g(x)$ can have a different shape near to $x = 0$ if the perturbation $g(x)$ contains terms of order lower than x^{k-1}. The larger the k value the more there are monomials whose presence in $g(x)$ may perturb the shape of a function near $x = 0$.

Noticeably, for the Morse function $V(x) = x^2$, the codimension is zero which corresponds to the structural stability of this function. Similarly, it will soon become apparent that the codimension of Morse functions in several variables is also equal to zero.

We shall now proceed to examples of the computation of the codimension of functions of two variables having a degenerate critical point.

The following example will demonstrate that not all functions have a finite codimension.

EXAMPLE A.4

$$V(x, y) = x^2 y$$

The Jacobian ideal of this function is given by

$$\Delta(V) = \{U_1(x, y)xy + U_2(x, y)x^2; U_1, U_2 \in E_2\}$$

To establish which monomials in two variables, contained in M_2, are absent from the Jacobian ideal $\Delta(V)$, we shall represent the monomials belonging to M_2 in the form of Pascal's triangle (this is a general procedure for computing the codimension of a function in two variables:

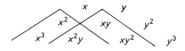

It follows from Pascal's triangle that the following monomials: x, y, y^2, y^3, ..., etc., are missing from $\Delta(V)$, and thus the function V has an infinite codimension.

As follows from the above example the function x^2y is not k-determined. Hence, it is not structurally stable, since each perturbation containing terms missing from $\Delta(V)$ changes the nature of its degenerate critical point. This is a more serious defect of the potential function than, for example, that of the function from Example A.3, which is k-determined and a universal unfolding containing a finite number of terms may be found for it.

The function x^2y may, however, be modified by adding a suitable term missing from $\Delta(V)$ in such a way that the new function is k-determined, see Example A.8.

EXAMPLE A.5

$$V(x, y) = x^3 + y^3$$

The function V is smooth and vanishes at $\mathbf{x} = (0, 0)$, so it belongs to the subset m_2 in the space E_2. In addition, the function has a degenerate critical point at $\mathbf{x} = (0, 0)$.

Indeed, $[\partial V/\partial x] = 0 = [\partial V/\partial y](0)$ and, moreover, its Hessian matrix (the matrix of second derivatives) determinant vanishes at $\mathbf{x} = (0, 0)$

$$\begin{vmatrix} \partial^2 V/\partial x^2 & \partial^2 V/\partial x \partial y \\ \partial^2 V/\partial y \partial x & \partial^2 V/\partial y^2 \end{vmatrix}_{(0, 0)} = \begin{vmatrix} 6x & 0 \\ 0 & 6y \end{vmatrix}_{(0, 0)} = 0$$

Let us now compute the Jacobian ideal of this function:

$$\Delta(V) = \{U_1(x, y)3x^2 + U_2(x, y)3y^2; U_1, U_2 \in E_2\}$$
$$= \{U_3 3x^2 + U_4 3y^2; U_3, U_4 \in E_2\}$$

To find out which monomials in two variables contained in M_2 are missing from the Jacobian ideal $\Delta(V)$, the monomials belonging to M_2 will be represented in the form of Pascal's triangle

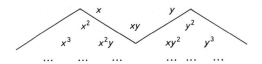

Since all monomials in the Jacobian ideal are of the form $U_3(x, y)x^2$, $U_4(x, y)y^2$, where U_3, U_4 are arbitrary monomials in two variables (1, x, y, etc.), the monomials belonging to M_2 and missing from $\Delta(V)$ are thus x, y, xy, as depicted in the above scheme. Consequently, the dimension of the set complementing Δ to M_3 is 3, $\text{cod}(V) = 3$.

Near to the critical point of the function $V(x, y) = x^3 + y^3$, see Example A.5, the derivatives $\partial V/\partial x = 3x^2$, $\partial V/\partial y = 3y^2$ are small and, accordingly, all terms occurring in the Jacobian ideal of this function are also small in the vicinity of the critical point $\mathbf{x} = (0, 0)$. In contrast, this cannot be said about the monomials x, y, xy. This leads to the following conclusions. Firstly, since terms of higher order are not missing from $\Delta(V)$, the function $V(x, y) = x^3 + y^3$ should be k-determined (the effect of perturbations containing monomials of higher order can be neglected). Secondly, the function V cannot be structurally stable, since the influence of perturbations containing the terms x, y, xy may not be disregarded. Thirdly, a universal unfolding of this function, that is a structurally stable family of functions containing the original function, should have the form

$$g(x, y; a, b, c) = ax + by + cxy$$

while the structurally stable family of functions, containing the original function is given by

$$V(x, y; a, b, c) = x^3 + y^3 + ax + by + cxy$$

The above example shows why a codimension is equal to the number of parameters in a universal unfolding. It follows from the Thom theorem (see Table 2.5) that the universal unfolding of the function $V(x, y) = x^3 + y^3$ is, indeed, expressed by the above equation.

EXAMPLE A.6

$$V(x, y) = x^3 + xy^3$$

The function V has a degenerate critical point at $\mathbf{x} = (0, 0)$; in addition, V is smooth and $V(0, 0) = 0$. Let us compute the Jacobian ideal of this function. It is a set given by the equation

$$\Delta(V) = \{U_1(x, y)(3x^2 + y^3) + U_2(x, y)(3xy^2); U_1, U_2 \in E_2\} \tag{A5}$$

as $\partial V/\partial x = 3x^2 + y^3$, $\partial V/\partial y = 3xy^2$.

Now, we have to find monomials from the space M_2 missing from $\Delta(V)$, i.e. those which cannot be represented in the form (A5); these will be monomials which do not have to be small near to the critical point of the function V.

We check that the following equalities hold:

$$x^3(1/3)x(3x^2 + y^2) + (-1/9)(3xy^2) \tag{A6}$$

$$xy^2 = 0(3x^2 + y^3) + (1/3)(3xy^2) \tag{A7}$$

$$y^5 = y^2(3x^2 + y^3) + (-x)(3xy^2) \tag{A8}$$

which means that all terms proportional to x^3, xy^2, y^5, belonging obviously to M_2, are also present in $\Delta(V)$. Additionally

$$x^2 = (1/3)(3x^2 + y^3) + (-1/3)y^3$$

that is the linear combination $[x^2 + (1/3)y^3]$ belongs to M_2. Hence, the dimension of space complementing Δ to M_2 is 7 — it contains the monomials x, x^2, xy, x^2y, y, y^2, y^3, y^4, the monomials x^2 and y^3 being linearly dependent. The terms from the spaces M_2 and Δ are conveniently represented in Pascal's triangle form,

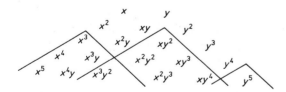

the linearly dependent monomials being printed in bold-face. The families of monomials belonging to Δ, generated by the monomials x_3, xy^2, y^5 (cf. equations (A6)–(A8)) are seen in the diagram.

EXAMPLE A.7

$$V(x, y) = x^5 + y^5$$

The Jacobian ideal of this function is of the form

$$\Delta(V) = \{U_1(x, y)x^4 + U_2(x, y)y^4; U_1, U_2 \in E_2\}$$

Proceeding now as in Example A.5 we establish that in the set complementing Δ to M_2 are the following monomials: $x^i y^i$, where $i, j = 1, 2, 3$. Hence, a universal unfolding of this function must contain the term $x^3 y^3$.

EXAMPLE A.8

$$V(x, y) = x^2 y + y^4$$

The Jacobian ideal of this function has the form

$$\Delta(V) = \{U_1(x, y)2xy + U_2(x, y)(x^2 + 4y^3); U_1, U_2 \in E_2\}$$

It follows from the form of Jacobian ideal and from the following equalities

$$x^3 = -4y^2(xy) + x(x^3 + 4y^3)$$
$$y^4 = -1/4 \, x(xy) + 1/4 \, y(x^2 + 4y^3)$$

that the Jacobian ideal contains al monomials in two variables proportional to xy, x^3, y^4

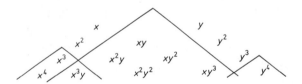

Seemingly, the Jacobian ideal does not contain five monomials: x, x^2, y, y^2, y^3. Since the monomials x^2, y^3 are linearly dependent:

$$x^2 = 0(xy) + 1(x^2 + 4y^3) - 4y^3$$

the codimension of the function $V(x, y) = x^2 y + y^4$ is equal to 4, and its universal unfolding has the form

$$V(x, y; a, b, c, d) = x^2 y + y^4 + ax + by + cy^2 + dy^3$$

according to the form of the family of potential functions D_2 given in Table 2.5, being reduced for $a = b = c = d = 0$ to the function $x^2 y + y^4$, having a degenerate critical point (k-determined).

A2.3 SUMMARY

Calculating the Jacobian ideal of the following functions in one and two variables having a degenerate critical point: x^3, $\pm x^4$, x^5, $\pm x^6$, x^7, $x^2 y - y^3$, $x^3 + y^3$, $\pm(x^2 y + y^4)$, $x^2 y + y^5$, $\pm(x^3 + y^4)$ according to the rules presented in Section A2.2 we conclude that the above functions are k-determined, because their codimension is finite. Furthermore, on the basis of the form of Jacobian ideal we establish that the respective universal unfoldings of these functions have a form consistent with the functions compiled in Tables 2.2, 2.5 (in Section A2.2 we computed the Jacobian ideal, among other functions, for the functions x, x^2, x^3, x^4, ..., and for $x^3 + y^3$, $x^2 y + y^4$).

Bibliographical Remarks

A good introduction to elementary catastrophe theory is a paper by Zeeman in *Scientific American*. A more systematic lecture on elementary catastrophe theory can be found in Gilmore's or Poston and Stewart's books, as well as in papers by Stewart and Zeeman. A paper by Poston and Woodcock, containing demonstrative graphs, can also be recommended.

References

W. I. Arnol'd, "Critical points of smooth functions", *Proc. Internat. Congr. Math.*, Vancouver, 1974, p. 19.

W. I. Arnol'd, "Critical points of smooth functions and their standard forms" (in Russian), *Usp. Matem. Nauk*, **30**, 603 (1975).

T. Bröcker and L. Lander, "Differentiable germs and catastrophes", *London Math. Soc. Lect. Notes*, **17**, London, 1975.

D. R. J. Chillingworth, "Elementary catastrophe theory", *Bull. Inst. Math. Applic.*, **11**, 155 (1975).

R. Gilmore, *Catastrophe Theory for Scientists and Engineers*, J. Wiley and Sons, New York, 1981.

M. Golubitsky and V. Guillemin, *Stable Mappings and their Singularities*, Springer-Verlag, New York, 1976.

J. Mather, "Stability of C-mappings. III. Finitely determined map germs". *Publ. Math. IHES*, **35**, 127 (1968).

J. Mather, "Stability of C-mappings. IV. Classification of stable germs by R-algebras", *Publ. Math. IHES*, **37**, 223 (1969).

T. Poston and I. N. Stewart, *Catastrophe Theory and its Applications*, Pitman, London, 1978.

I. N. Stewart, "The seven elementary catastrophes", *New Scientist*, **68**, 447 (1975).

I. N. Stewart, "Applications of catastrophe theory to physical sciences", *Physica*, **20**, (2), 245 (1981).

I. N. Stewart, "Catastrophe theory in physics", *Repts. Progr. Phys.*, **45** (2), 185 (1982).

R. Thom, *Structural Stability and Morphogenesis*, Benjamin-Addison-Wesley, New York, 1975.

D. Trotman and N. C. Zeeman, "Classification of elementary catastrophes of codimension 5" *Lect. Notes in Math.*, **525**, Springer-Verlag, Berlin, 1976.

A. Woodcock and T. Poston, "Geometrical study of the elementary catastrophes", *Lect. Notes in Math.*, **373**, Springer-Verlag, Berlin, 1974.

E. C. Zeeman, "Catastrophe theory", *Scient. Amer.*, **234**, 65 (1976).

E. C. Zeeman, *Catastrophe Theory*, Pitman, London, 1978.

E. C. Zeeman, "Bifurcation and catastrophe theory", *Contemporary Math.*, **9**, 207 (1982).

Chapter 3

Applications of Elementary Catastrophe Theory (Non-Chemical Systems)

3.1 INTRODUCTION

In Section 1.3 we described systems in which qualitative and discontinuous state change taking place as a result of a continuous change in control parameters — that is catastrophes, can be observed. These were the following systems:

(1) a soap film,
(2) the liquid–vapour system,
(3) a beam of radiation falling upon a lens,
(4) an insect population,
(5) the heartbeat,
(6) chemical reactions,
(7) a non-linear recurrent equation.

In this chapter we shall show how the observed phenomena may be explained by means of elementary catastrophe theory. In principle, the discussion will be confined to examination of non-chemical systems. However, some of the discussed problems, such as a stability of soap films, a phase transition in the liquid–vapour system, diffraction phenomena or even non-linear recurrent equations, are closely related to chemical problems. This topic will be dealt with in some detail in the last section. The discussion of catastrophes (static and dynamic) occurring in chemical systems is postponed to Chapters 5, 6; these will be preceded by Chapter 4, where the elements of chemical kinetics necessary for our purposes will be discussed.

3.2 STABILITY OF SOAP FILMS

A simple example of a physical system exhibiting in certain circumstances sensitivity to arbitrarily small changes in control parameters is a soap film stretched on a thin wire frame. We shall demonstrate that such a system can be described by a potential function and that an elementary catastrophe corresponding to this potential function may occur in such a system.

We shall examine properties of two systems of this type. Let the first system consist of a soap film stretched on two rings having the same diameter R (Fig. 28).

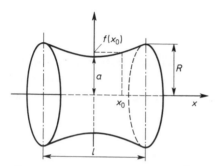

Fig. 28. Parametrization of the surface of a soap film.

In our considerations we shall neglect the effect of gravity forces on the film and the effects associated with a non-zero thickness of the film. Under these assumptions the potential energy of the film is equal to the potential energy of the surface tension forces. The shape of a soap film stretched on an arbitrary contour is thus determined by the acting forces of the surface tension. The surface formed must minimize the potential energy of the surface tension forces. As the potential energy of the surface tension forces in proportional to the surface area of the film, the soap film takes the shape minimizing the area of the surface stretched on a given contour.

The system described above belongs to a class of models with potential. The properties of stationary states of the system are totally determined by the condition of a minimum of potential energy of the system. In this case, the potential energy is proportional to the surface area of the film. Since the gravitational field is a potential field, accounting for the effect of gravitational forces on the film also leads to a certain model with potential.

In the considered case the soap film takes the shape of a surface of revolution, see Fig. 28. The area of revolution, S, is given by the equation

$$S = 2\pi \int_{-l/2}^{l/2} f(x)[1 + f'^2(x)]^{1/2} \, dx \qquad (3.1)$$

where $f(x)$ is the distance of the film from the axis of symmetry at x (see Fig. 28). Hence, the stationary states of the system correspond to the condition $S = \text{min}$. Let us assume that R is fixed. Then, a control parameter is the distance between the rings l, a state function is $f(x)$ and a state equation is (3.1) with the condition $S = \text{minimum}$. The function $f(x)$ must obey the boundary conditions

$$f(1/2) = f(-1/2) = R \qquad (3.2a)$$

When l is not too large (see further analysis), a solution to the equation of state meeting the boundary conditions (3.2a) is given by

$$f(x) = s\cosh(x/a), \qquad a \equiv f(0) \qquad (3.2b)$$

and it follows from (3.2b) that the boundary conditions (3.2a) may be represented in the form

$$e^{1/2a} + e^{-1/2a} = 2R/a \qquad (3.2c)$$

Solution (3.2) is a so-called catenary curve; such a shape is taken by a freely sagging chain and the surface of revolution, see Fig. 28, is called a catenoid. The value of parameter a can be computed for a given l from the boundary condition (3.2c). The parameter a fully determines properties of the solution (3.2b) and may therefore be considered to be a function of state.

Even a cursory examination of equations (3.2) reveals that the function (3.2b) cannot be a solution of the variation problem, $S = \text{min}$, for all values of the control parameter l. Namely, for set values of R, l equation (3.2c) may not have solutions for the parameter a.

Indeed, equation (3.2c) may be written as

$$w(r) = h(r), \quad w(r) \equiv e^{\lambda r/2} + e^{-\lambda r/2}, \quad h(r) \equiv 2r, \quad r \equiv R/a, \quad \lambda \equiv l/R \ (3.3)$$

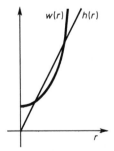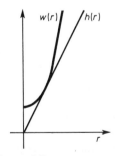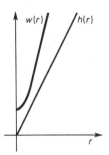

Fig. 29. Functions $w(r) = e^{\lambda r/2} + e^{-\lambda r/2}$, $h(r) = 2r$.

Three possible cases for the equality $w(r) = h(r)$ are shown in Fig. 29.

In a critical (limit) case, the curves $w(r)$ and $h(r)$ are tangential, i.e. their first derivatives are equal for some value of r. Hence, the critical values are calculated by solving the system of equations

$$w(r) = h(r), \text{ that is } e^{\lambda r/2} + e^{-\lambda r/2} = 2r \tag{3.4a}$$

$$w'(r) = h'(r), \text{ that is } (\lambda/2)(e^{\lambda r/2} - e^{-\lambda r/2}) = 2 \tag{3.4b}$$

Dividing the first of the equations by the second we obtain

$$(\lambda r/2) = \operatorname{ctgh}(\lambda r/2) \tag{3.5}$$

On computing from (3.5) $t_{cr} \equiv \lambda_{cr} r_{cr}/2$ we may determine from (3.4a) the value of r_{cr} and λ_{cr} on the basis of knowledge of t_{cr}. Solving the non-linear equation (3.5), $t_{cr} = \operatorname{ctgh}(t_{cr})$, yields $t_{cr} \cong 1.1997$ and it follows from (3.4a) and from the definition of t_{cr} that $\lambda_{cr} \cong 1.3255$, $r_{cr} \cong 1.8102$.

Hence, when the control parameter l increases to such a value that the ratio $l/R \equiv \lambda$ exceeds the critical value 1.3255, then the boundary conditions (3.2c) will not be met and the function (3.2b) ceases to be a solution for the minimum surface (minimum of the potential energy). The critical value of a is $a_{cr} = R/1.8102$.

Racapitulating on our considerations we may state that a catastrophe will occur in the system when the control parameter l exceeds the critical value $l_{cr} \cong 1.3255\,R$. As the function (3.2b) will not be then a solution to the variation problem $S = \text{minimum}$, the soap film has to take some other shape.

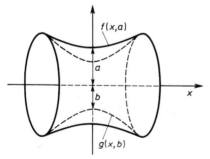

Fig. 30. Family of surfaces of revolution $g(x;\, b,\, c) = b\cosh(x/c)$.

We shall now examine in more detail an evolution of the shape of a soap film and a catastrophe in this system on changing the control parameter l.

For the purpose, let us consider a family of surface of revolution, Fig. 30, for which the distance from the axis of rotation x is given by

$$g(x; b, c) = b\cosh(x/c) \tag{3.6}$$

where b, c are some parameters. In the special case $b = c = a$ equation (3.6) defines the examined surface of revolution for which the distance from the axis of rotation is given by (3.2b).

The family of functions of revolution defined by equation (3.6) may be regarded as a set of surfaces obtained by symmetrical deformation (with preservation of rotational symmetry) of the initial surface (3.2). The introduced family of surfaces satisfies the following conditions for the distance from the axis of rotation

$$g(0, b, c) = b \tag{3.7a}$$

$$g(\pm l/2, b, c) = b\cosh(l/2c) = R \tag{3.7b}$$

Hence, the parameter b corresponds to the distance of the surface from the axis of rotation, while the condition (3.7b) for a given $b < R$ may always be met by a suitable choice of c (the parameter c thus depends on the parameter b). Firstly, equation (3.7b) may be written in the form of a quadratic equation

$$z^2 - (2R/b)z + 1 = 0, \qquad z = e^{1/2c} \tag{3.8}$$

which has a positive solution for $b < R$ (meaning of the condition $b < R$ follows from Fig. 30). Secondly, the condition (3.7) may be written in the form analogous with equation (3.3)

$$w(r) = h(r), \qquad w(r) \equiv e^{\lambda r/2\,(b/c)} + e^{-\lambda r/2\,(b/c)}, \qquad h(r) \equiv 2r$$

$$r \equiv R/a, \qquad \lambda \equiv l/R \tag{3.9}$$

Repetition of considerations presented in equations (3.4), (3.5) leads to an equation analogous with (3.5)

$$(\lambda r/2)(b/c) = \operatorname{ctgh}\left[(\lambda r/2)(b/c)\right] \tag{3.10}$$

It is now evident that for a given b value one may always select such a value of the parameter c that equation (3.10) is valid.

We shall now examine the process of continuous changes in the system defined in Fig. 28, generated by a continuous variation of the control

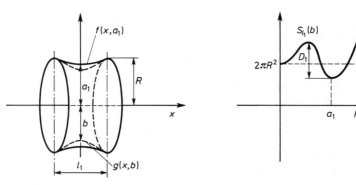

Fig. 31. Surface of a soap film for $l = l_1$.

parameter l. For small $l = 1$, we have the following situation depicted in Fig. 31, where $S(b)$ stands for the surface area calculated from equation (3.1)

$$S(b) = 2\pi \int_{-l/2}^{l/2} g(x; b, c)\left[1 + g'^2(x; b, c)\right]^{1/2} dx \qquad (3.11)$$

and from equations (3.6), (3.7b) for a given b. The parameter b may thus be considered to be a state variable of the system.

Seemingly, the surface of revolution set by the function $g(x; b, c) = b\cosh(x/c)$, $b = c = a_1$, and a_1 meets the condition (3.2c), i.e. has the smallest surface area among all surfaces of revolution set by the family of functions (3.6).

Hence, the solution (3.2b) is stable — a deformed soap film returns to the initial shape if the perturbation energy does not exceed the value D_1 (the energy is expressed in terms of area units).

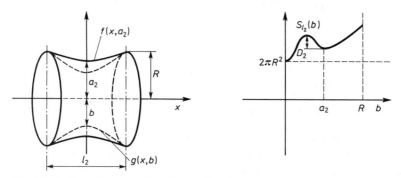

Fig. 32. Surface of a soap film for $l = l_2 > l_1$.

For a suitably larger l, $l = l_2 > 1$, the dependence of surface area on b is depicted in Fig. 32.

The surface of revolution defined by the function of distance from the rotation axis $g(x; a_2, a_2) = a_2 \cosh(x/a_2) \equiv f(x; a_2)$ corresponds to a local minimum of the function $S(b)$ computed on the basis of equations (3.7b), (3.11) for the family of functions (3.6).

The solution for $l = l_2$, $f(x; a_2)$ is thus stable if the perturbation energy does not exceed D_2 — on removing the perturbing force the deformed film will return to the initial state.

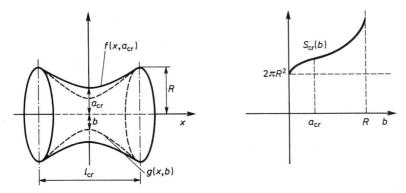

Fig. 33. Surface of a soap film for $l = l_{cr}$.

For a still larger distance between the rings, the following plot is obtained (this is the critical distance $l = l_{cr}$) (Fig. 33), where a_{cr} corresponds to a point of inflection of the curve $S(b)$. Hence, the surface of revolution for $l = l_{cr}$, for which the distance from the rotation axis is given (3.2b), (3.2c), is unstable: an arbitrarily small deformation of the film resulting in narrowing

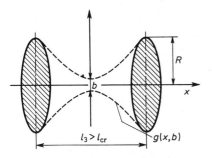
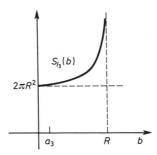
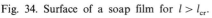

Fig. 34. Surface of a soap film for $l > l_{cr}$.

of the surface leads to the surface having a smaller area and, consequently, lower potential energy. Thus perturbed film further narrows until it breaks and jumps on the side surfaces of the rings. Such a final state is stable, since it has the minimum surface area equal to $2\pi R^2$.

Hence, on further increasing l, the function (3.2b) is no longer a solution to the variation problem $S =$ minimum and the only stable configuration of a film stretched on two rings separated by $l > l_{cr}$ are two side surfaces of the rings, see Fig. 34.

The process of increasing the control parameter l to the value $l = l_{cr}$ leads to an attainment by the system of a state sensitive to arbitrarily small perturbations. Thus, at the point $l = l_{cr}$ a catastrophe takes place; the system exhibitis a discontinuity in evolution with a continuous change of parameter.

It follows from the form of the function $S_l(b)$, see Figs. 30–34, that it may be parametrized in the following manner:

$$S_l(b) = \text{const} \left[(b - b_{cr})^3 + (l - l_{cr})(b - b_{cr}) \right] \tag{3.12}$$

The function $S(b; l) \equiv S_l(b)$, playing the role of a potential function of the system, thus corresponds to the potential function of an elementary fold catastrophe (A_2), b plays the role of a state variable and l is a control parameter.

In the system consisting of a film stretched on a wire contour, more complicated elementary catastrophes may occur upon varying suitable control parameters. In the cited books by Fomyenko, a swallowtail catastrophe (A_4) of a soap film has been described.

3.3 THERMODYNAMICS AND PHASE TRANSITIONS

A classical theory of phase transitions may be formulated by means of elementary catastrophe theory. We shall describe some notions of the theory of phase transitions in terms of elementary catastrophe theory. Next, we shall describe examples of application of catastrophe theory to the description of the liquid–vapour equilibrium.

3.3.1 Classification of phase transitions

Lest us assume that the physical system being considered may be described by means of a family of potential functions $U(\mathbf{x}; \mathbf{c})$, depending on

n state variables $\mathbf{x} = (x_1, ..., x_n)$ and k parameters $\mathbf{c} = (c_1, ..., c_k)$. Such a potential function may be a thermodynamic function, for example the Gibbs function (the free energy). The conditon for thermodynamic equilibrium is a minimum of a thermodynamic function U. A continuous variation of control parameters $\mathbf{c}$ may result in a change in local properties of the potential U — a catastrophe, which can be related to a certain phase transition.

On the basis of analysis of experimental facts Ehrenfest has introduced the following classification of phase transitions: a conversion is called the phase transition of nth order if successive derivatives of a thermodynamic function U up to and including $(n - 1)$ are continuous functions, whereas the nth derivative has a step discontinuity at the transition point: the

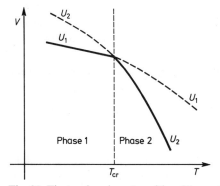

Fig. 35. First-order phase transition (V — volume, T — temperature, p — pressure, $p = $ const).

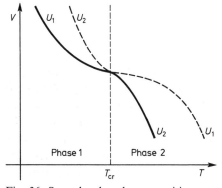

Fig. 36. Second-order phase transition, $p = $ const.

left-hand and right-hand derivatives have different values at the transition point. According to this definition the potential U and its derivatives will have the following shape with respect to each independent variable: at the first-order transition the function U is continuous with a fold at the transition point so that the first derivative has a discontinuity (Fig. 35) while at the second-order transition the function U is continuous and smooth, its first derivative is continuous and folded, whereas the second derivative is discontinuous (Fig. 36).

The Ehrenfest classification is not too well suited for the description of real phase transitions occurring in nature. The above remark concerns most of all the phase transitions which are not first-order. Better suited for an examination and classification of phase transitions is the Landau classification. Landau's idea is based on an assumption that in the case of many phase transitions one may always find a quantity, called the parameter of order, whose small change (with respect to the value $\eta = 0$) causes a qualitative changes in the parameters of a body (this implies that for $\eta = 0$ the system is in the sensitive state).

Landau assumed that near the phase transition a thermodynamic function, dependent on the order parameter η and on some other variables X, may be expanded in a power series in the order parameter η

$$U(X;\eta) = U(X;0) + \left[(\partial U/\partial \eta)(0) \right] \eta + \ldots \tag{3.13}$$

At equilibrium, the condition $(\partial U/\partial \eta)(0) = 0$ must hold. According to the Landau concept, qualitative changes in a system take place when the parameter η passes through zero; hence, the critical point of potential (with respect to η) must be a singular point for the state $\eta = 0$ to be sensitive.

3.3.2 The liquid-vapour phase transition

The liquid-vapour phase transition is a first-order transition. For example, the molar entropy S, being a derivative of te Gibbs potential $G, S = -(\partial G/\partial T)_p$, where T is the temperature and p the pressure, has a discontinuity at the boiling point $T = T_b$. For a description of the liquid–vapour equilibrium, various approximate state equations are employed. Let us consider one of the most common equations of this type, the semiempirical van der Waals equation, written in terms of temperature T, volume V and pressure p or a gas

$$(p + \alpha/V^2)(V - \beta) = RT \tag{3.14}$$

where R is the gas constant and the constants α and β depend on the kind of gas. The constant α is related to interactions between atoms or molecules of a gas and the constant β is associated with the size of atoms or molecules. A typical isotherm obtained from (3.14) is depicted in Fig. 37.

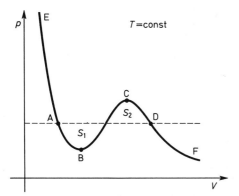

Fig. 37. The van der Waals isotherm.

In a real liquid–vapour system the following relationship must hold:

$$(\partial p/\partial V)_T \leqslant 0 \tag{3.15a}$$

This inequality, called the condition of mechanical stability, implies that if at a constant temperature the pressure in a system increases, then its volume decreases. It follows from Fig. 37 that along the section BC the van der Waals isotherm does not satisfy the condition (3.15a). This signifies that the isotherms obtaind from equation (3.14) are, at least partially, non-physical, and the corresponding states are physically unattainable. Thus, equation (3.14) is only an approximate state equation.

The non-physical van der Waals isotherm may be improved using the so-called Maxwell construction. It involves drawing the horizontal section AD, for which $(\partial p/\partial V)_T = 0$, joining the two branches of the isotherm, EA and DF, corresponding to the liquid and gaseous phase of a system, respectively. It follows from the condition of equality of chemical potentials at a critical point that the section AD should be thus selected that the areas S_1 and S_2 be equal. Between the points A and D the system is nonhomogeneous, i.e. separated into two phases coexisting in equilibrium. The sections AB and CD of the van der Waals isotherm can be realized experimentally and correspond to the states of the superheated liquid and those of the supercooled vapour.

Figure 38 illustrates the family of isotherms of equation (3.14) plotted using the Maxwell construction.

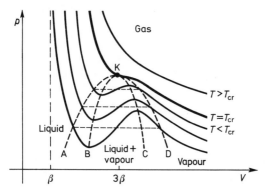

Fig. 38. The Maxwell construction for the van der Waals isotherms.

The locus of endpoints of physical branches of the isotherms is the curve BKC bounding the region in which the system cannot be one-phase. At the same time, this curve determines the applicability region of the van der Waals equation. The phase equilibrium curve AKD, bounding the region in which two phases coexist in equilibrium, is obtained in a similar way. In the region between the curves AKD nad BKC, metastable states of the superheated liquid or those of the supercooled vapour may occur. The point K, at which the curves AKD and BKC are tangential, i.e. the point having coordinates (p_{cr}, V_{cr}, T_{cr}), is called the critical point. The critical point can be defined as a point of inflection of the critical isotherm

$$(\partial p/\partial V)_T = 0 \tag{3.15b}$$

$$(\partial^2 p/\partial V^2)_T = 0 \tag{3.15c}$$

Hence, it is a degenerate critical point meeting the above conditions (see Section 2.2.2).

At the critical point the difference between the coexisting phases disappears. The difference between the fluid density and its critical density (density at the critical point) may be adopted as an order parameter.

3.3.3 Application of elementary catastrophe theory to a description of the liquid-vapour transition

Let us now return to the van der Waals equation (3.14). The equation is non-linear in V. The presence of third power of V and the requirement of vanishing of the first and second derivatives suggest that in some circumstances a cusp catastrophe (A_3) may appear in the system. Hence, a change of variables should be performed in equation (3.14) to reduce, if possible, the equation to the form resulting from a structurally stable potential function of the cusp catastrophe $V = x^4 + ax^2 + bx$.

We shall examine local properties of the liquid–vapour system near the critical point (p_{cr}, V_{cr}, T_{cr}) whose coordinates result directly from equations (3.14), (3.15b), (3.15c):

$$(p_{cr}, V_{cr}, T_{cr}) = [\alpha/(27\,\beta^2),\, 3\beta,\, 8\alpha/(27\,\beta R)] \tag{3.16a}$$

On carrying out normalization of variables

$$p' = p/p_{cr}, \qquad V' = V/V_{cr}, \qquad T' = T/T_{cr} \tag{3.17}$$

we obtain the so-called reduced van der Waals equation

$$(p' + 3/V'^2)(V' - 1/3) = 8/3\ T' \tag{3.18}$$

the critical point now having the coordinates

$$(p_{cr}', V_{cr}', T_{cr}') = (1, 1, 1) \tag{3.16b}$$

Note that equations (3.18), (3.16b) do not contain the constants α, β, characteristic of a given gas. Thus, it is an equation of state for all the liquid-vapour systems to which the van der Waals equation may be applied.

Two systems whose reduced variables p', V', T' have the same values, are said to be in corresponding states. The reduced isotherms deriving from equation (3.18) are identical for all gases (the law of corresponding states).

We shall now perform the next change of variables, switching from volume to density $X' = 1/V'$ (for unit mass), and obtaining from (3.18) a new equation

$$(p' + 3X'^2)(1/X' - 1/3) = 8/3\ T' \tag{3.19}$$

Finally, upon shifting the critical point $(p_{cr}', x_{cr}', T_{cr}') = (1, 1, 1)$ to the origin (all functions of elementary catastrophes have a degenerate critical point at the origin):

$$P = p' - 1, \quad x = X' - 1, \quad t = T' - 1 \tag{3.20}$$

$$\left(P_{\text{cr}}, x_{\text{cr}}, t_{\text{cr}}\right) = \left(0, 0, 0\right) \tag{3.16c}$$

we obtain the equation of state

$$x^3 + ax + b = 0$$

$$a = \left(8t + P\right)/3, \quad b = \left(8t - 2P\right)/3 \tag{3.21}$$

fully equivalent to the initial van der Waals equation (3.14).

Equation (3.21) derives from the requirement of a minimum of the potential function, $dU(x; a, \text{b})/dx = 0$

$$U(x; a, b) = 1/4\,x^4 + 1/2\,ax^2 + bx \tag{3.22}$$

the potential U being a function of density, temperature and pressure. Let us add that in equation (3.22) x plays the role of a parameter of order (see end of Section 3.3.2).

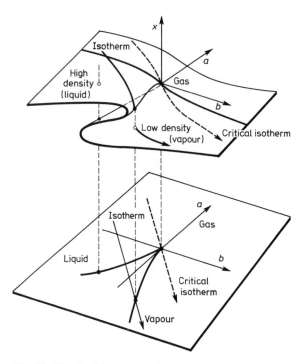

Fig. 39. The liquid–vapour phase transition on the surface of the cusp catastrophe (A_3).

In summary, the van der Waals equation may be derived from the condition of a minimum of a potential function corresponding to the cusp catastrophe (A_3) in which the state variable x is the fluid (liquid or vapour) density, $x = V_{cr}/V - 1$, while the parameters a, b are functions of the temperature T and pressure p (see equation (3.21)).

The first important conclusion which can be drawn from the above construction is a structural stability of phenomena described by the van der Waals equation. The conclusion is consistent with an observation that the reduced van der Waals equation (3.18) describes the critical transition for a number of gases. The liquid–vapour phase transition may thus be described on the surface of the cusp catastrophe, see Fig. 39, in which a projection on the control parameters plane and the isotherm $T = $ const are also shown.

The analysis of behaviour of the system on the surface of the A_3 catastrophe reveals that some fragments of the surface are unavailable during an evolution of the system controlled by continuous changes in control parameters a, b (or T, p). Such unavailable regions were marked with a broken line on the section of the catastrophe surface with the plane $2a + b = $ const (i.e. $t = $ const), Fig. 40.

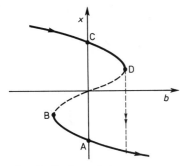

Fig. 40. Section of the cusp catastrophe surface for the liquid–vapour system.

The above observation has a physical interpretation. The unavailable region BD of the isotherm, at a continuous evolution on the catastrophe surface, corresponds to the region bounded by the curve BKC in Fig. 38. The sections AB and CD correspond to the superheated liquid and the supercooled vapour, respectively. It should be emphasized that the unavailability of the states BD derives from a model of the A_3 catastrophe provided that the system evolution takes place on the catastrophe surface M_3 at a continuous variation of control parameters.

A family of potential functions $U(x; a, b)$ is a fourth-order polynomial. On changing, for example, the parameter a the plot of this polynomial varies in the way typical for a first-order phase transition (Fig. 41).

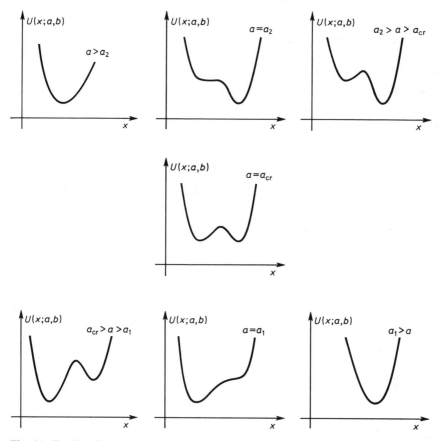

Fig. 41. Family of potential functions $U(x; a, b) = x^4 + ax^2 + bx$, $b = $ const.

The bifurcation set B_3 is given by the condition $4a^3 + 27b^2 = 0$, which may be expressed in terms of pressure p and temperature T

$$4a^3 + 27b^2 = 4[(8t + P)/3]^3 + 27[8t - P)/3]^2 = 0$$

$$t \equiv 28\beta RT/(8\alpha) - 1, \qquad P \equiv 27\beta^2 p/\alpha - 1 \qquad (3.23)$$

The above equation allows us to determine the values of the control parameters, p, T, for which a catastrophe, i.e. a spontaneous transition from the state of the superheated liquid to that of the vapour or from the state of the supercooled vapour to that of the liquid, may occur. Let us note at this point that the formalism of elementary catastrophe theory does not permit a description of evolution of the system on leaving the catastrophe surface until returning onto this surface.

The most singular point of a cusp catastrophe occurs at the origin (this convention holds for every elementary catastrophe). This is a critical point having the coordinates (x, a, b) or $(p, V, T) = (p_{cr}, V_{cr}, T_{cr})$. From the mathematical viewpoint, it corresponds to a degenerate critical point of the potential function $U(x; 0, 0) = 1/4\, x^4$, the plot of the function $dU/dx = x^3$ being marked in Fig. 39 with a broken line. If a system controlled by the control parameters a, b (p, T) may remain in the immediate neighbourhood of the point $(0, 0, 0)$ for an extended time, then on the occurrence of a catastrophe the system may return to the set Σ_3 due to only a small change in control parameters and a subsequent catastrophe, accompanied by a change in the fluid density x, may take place. The system being in the vicinity of a critical point, is particularly sensitive to fluctuations in control parameters (the sensitive state). Such a trajectory of the system near the origin in the control parameters space is shown in Fig. 42.

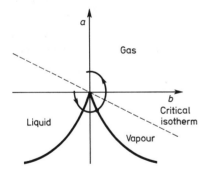

Fig. 42. Evolution of the liquid–vapour system near the critical point in the parameter space.

When the system is in the neighbourhood of the critical point, interesting experimental effects related to light scattering by such a system occurring in the phase transition region may be observed. As follows from the previous considerations, density fluctuations may appear in the liquid–vapour system being in the vicinity of the critical point. Media, being optically transparent under standard conditions, become opaque near the critical point. Intensity of radiation scattered at an angle with respect to the incident wave increases rapidly when $T \rightarrow T_{cr}$. The phenomenon is called critical opalescence. The temperature dependence is depicted in Fig. 43.

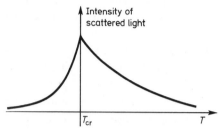

Fig. 43. Intensity of light scattered at a constant angle by the liquid–vapour system nearby the critical point.

The phenomenon of critical opalescence has been first elucidated by Smoluchowski who demonstrated that the increase in intensity of radiation scattered at a given angle results from the density fluctuations. In states far from the critical point the effect of fluctuations in control parameters on the properties of a system may be ignored. When the temperature approaches critical value, the sensitivity of the system to perturbations grows and local fluctuations in density in the system rapidly increase. As a result, the system becomes spatially nonhomogeneous. Fluctuations in the fluid density near the critical point may be regarded as the incipience of a new phase. The critical opalescence results from light scattering on local density fluctuations.

3.4 DIFFRACTION CATASTROPHES

3.4.1 Introduction

The next field of applications of elementary catastrophe theory are optical and quantum diffraction phenomena. In the description of short wave phenomena, such as propagation of electromagnetic waves, water waves, collisions of atoms and molecules or molecular photodissociation, a number of physical quantities occurring in a theoretical formulation of the phenomenon may be represented, using the principle of superposition, by the integral

$$\Phi(\mathbf{c}) = (k/2\pi)^{n/2} \int a(\mathbf{x}; \mathbf{c}) \exp\left[ikf(\mathbf{x}; \mathbf{c})\right] d^n\mathbf{x} \tag{3.24}$$

where k is the wavenumber, associated with the wavelength λ via the relation $k = 2\pi/\lambda$; the amplitude $a(\mathbf{x}; \mathbf{c})$ and phase $f(\mathbf{x}; \mathbf{c})$ are real and slow varying functions.

The meaning of variables $\mathbf{x} = (x_1, ..., x_n)$, $\mathbf{c} = (c_1, ..., c_k)$ will be exemplified by an electromagnetic wave propagating in a medium from one two-dimensional surface to another (this is the case $n = 2$, $k = 2$). The vector $\mathbf{x} = (x_1, x_2)$ describes the position of a source of signal on the first surface and the vector $\mathbf{c} = (c_1, c_2)$ is the position of a point of observation on the second surface. Then, an intensity of the signal at the point (c_1, c_2) is equal to the square of the absolute value of the function (3.24) for $n = 2$, $k = 2$, $a(\mathbf{x}; \mathbf{c})$ being the amplitude of a signal transmitted at the point (x_1, x_2) on the first surface and observed at the point (c_1, c_2) on the second surface, while the function $f(\mathbf{x}; \mathbf{c})$ is related to optical properties of the medium (to the refractive index).

Parameters $\mathbf{c}$ determining the position of a screen may be readily varied during the observations and are therefore conveniently regarded as control parameters; on the other hand, $\mathbf{x}$ will be considered as state variables. In the short-wave limit (so-called geometrical optics approximation) $\lambda \ll 1$ (or $k \gg 1$), the integrand in equation (3.24) rapidly oscillates. It will be shown in Section 3.4.2 that there is practically no contribution to the integral (3.24) from the points $\mathbf{x}$ at which the integrand strongly oscillates. This is apparent for a function in one variable — the integral of the function $a(x)\cos\left[kf(x)\right]$ is small for large k and slow varying functions $a(x)$, $f(x)$ due to cancelling of contributions to this integral from the regions where $\cos(kf)$ has the opposite sign, see Fig. 44 below.

A function is known to vary most slowly in the vicinity of a minimum or

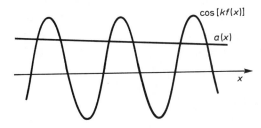

Fig. 44. Slow varying function $a(x)$ and fast oscillating function $\cos[kf(x)]$, $k \gg 1$, $f(x)$ — slow varying function.

a maximum. It follows from the above considerations that the main contribution to the integral (3.24) comes from the point $\mathbf{x}$ at which the phase $kf(\mathbf{x}; \mathbf{c})$ has an extremum (a critical point), $\nabla_{\mathbf{x}} f(\mathbf{x}; \mathbf{c}) = 0$ (cf. equation (2.1a)).

The relation with elementary catastrophe theory is now apparent: the function $f(\mathbf{x}; \mathbf{c})$ will play the role of a potential function, whereas critical points of this function determine, as will be shown in Section 3.4.2, the diffraction pattern (just as in the case of a classical potential function its critical points determine the stationary states of a system).

In many cases the integral (3.24) may be expressed in terms of oscillatory integrals of a simpler type

$$\Phi(\mathbf{c}) = (k/2\pi)^{n/2} \int \exp(ikF(\mathbf{x}; \mathbf{c})) \, d^n x \qquad (3.25)$$

the function $F(\mathbf{x}; \mathbf{c})$ being directly related to one of the functions of Thom's elementary catastrophe (see Chapter 2). Such integrals will be referred to as the canonical diffraction (oscillatory) integrals.

Integrals of the type (3.25), directly related to potential functions of elementary catastrophes, appear from attempts to describe the optical and quantum phenomena in the short-wave approximation, $\lambda \to 0$ or $k \to \infty$. The methods associated with the short-wave limit in optics are sometimes called catastrophe optics, since in the limit $k \to \infty$ straightforward methods of computation of the integrals (3.24) frequently fail. For example, at the points of coincidence of trajectories of two or more rays of light, that is on caustics — sites of higher intensity of the electromagnetic field (Greek kaustikos — capable of burning), straightforward estimations of the integrals (3.24) lead to divergence, yielding an infinite intensity of light.

In a later passage of the chapter we shall consider the short-wave limit for the Helmholtz equation, describing optical phenomena, and for the Schrödinger equation, describing quantum phenomena. The relation between the obtained oscillation integrals and the elementary catastrophe functions will be revealed and straightforward examples of application of the canonical diffraction integrals (3.25) discussed.

3.4.2 The short-wave limit for the Helmholtz equation

Optical phenomena may be described in a variety of cases by a scalar equation called the Helmholtz equation

$$\nabla^2 \varphi(\mathbf{x}) + k^2 n^2(\mathbf{x})\varphi(\mathbf{x}) = 0 \tag{3.26}$$

where n is the refractive index of a medium and k is the wavenumber ($k = 2\pi/\lambda$, where λ is the wavelength), which may be regarded as an approximation of the Maxwell equations.

To examine the short-wave limit, $k \to \infty$, for the Helmholtz equation, the function $\varphi(\mathbf{x})$ will be represented as

$$\varphi(\mathbf{x}) = a(\mathbf{x})\exp[ikf(\mathbf{x})] \tag{3.27}$$

where $a(\mathbf{x})$ and $f(\mathbf{x})$ are the amplitude and the phase, respectively, both the functions being real. Substitution of relation (3.27) into equation (3.26) yields, on separating the equation into the real and imaginary terms, the following equations

$$\nabla f \cdot \nabla f = n^2 + \nabla^2 a/(k^2 a) \tag{3.28a}$$

$$\nabla \cdot (a^2 \nabla f) = 0 \tag{3.28b}$$

a dot denoting the scalar product of two vectors, $\mathbf{A} \cdot \mathbf{B} = A_1 B_1 + ... + A_n B_n$. Presumably, for large k, $k \gg 1$, the term $\nabla^2 a/(k^2 a)$ in equation (3.28a) may be neglected. We shall then obtain, as a first approximation,

$$|\nabla f|^2 \equiv \nabla f \cdot \nabla f = n^2 \tag{3.29a}$$

$$\nabla f(a^2 \nabla f) = 0 \tag{3.29b}$$

Equation (3.29a) contains only f; hence, it enables the determination of $|\nabla f|$, $|\nabla f| = n$. On the other hand, the solution to equation (3.29b) will be found by carrying integration over a thin beam of rays passing near the points $\mathbf{x}_0$ and $\mathbf{x}$ (Fig. 45).

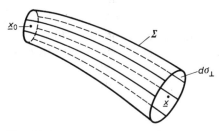

Fig. 45. Beam of rays having directions ∇f.

A beam of rays defines a three-dimensional region V, whose lateral surface is determined by curves such that their tangents coincide at each point with ∇f. The region V is bounded by surfaces perpendicular to ray directions. In a special case $n = $ const, the rays are parallel. Using equation (3.29b), the Gauss theorem for the transformation of an integral over the volume V into an integral over the surface Σ bounding V, followed by equation (3.29a), we obtain

$$0 = \int_v \nabla \cdot (a^2 \nabla f) dV = \int_\Sigma a^2 \nabla f \cdot d\boldsymbol{\sigma} = \int_\Sigma a^2 |\nabla f| \cos(\gamma) d\sigma = \int_\Sigma a^2 n d\sigma_\perp$$

$$= a^2(\mathbf{x}_0) n(\mathbf{x}_0) d\sigma_\perp(\mathbf{x}_0) - a^2(\mathbf{x}) n(\mathbf{x}) d\sigma_\perp(\mathbf{x}) \tag{3.30}$$

where γ is the angle between the normal to the surface Σ and the direction ∇f. The last equality in (3.30) derives from the method of construction of the region V (parallelism of a lateral part of the surface Σ to ∇f except for those fragments $d\sigma_\perp$ which are perpendicular to ∇f).

Eventually, an approximate relationship

$$\varphi(\mathbf{x}) = a(\mathbf{x}) \exp[ikf(\mathbf{x})]$$

$$|\nabla f(\mathbf{x})| = n(\mathbf{x}) \tag{3.31a}$$

$$a(\mathbf{x}) = a(\mathbf{x}_0) \left[\frac{n(\mathbf{x}_0) d\sigma_\perp(\mathbf{x}_0)}{n(\mathbf{x}) d\sigma_\perp(\mathbf{x})} \right]^{1/2} \tag{3.31b}$$

is obtained.

The asymptotic equation (3.31) has a serious defect: it is divergent when the denominator in (3.31b) vanishes which results in an infinite light intensity at such a point.

The reason for the divergence is obvious: $a(\mathbf{x})$ cannot be determined from (3.30) if

$$n(\mathbf{x}) d\sigma_\perp(\mathbf{x}) = 0 \tag{3.32}$$

Unfortunately, the expression (3.32) vanishes at most important (from the physical standpoint) points since $d\sigma_\perp$ vanishes as the points through

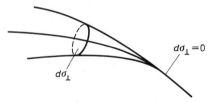

Fig. 46. Rays forming the caustic.

which pass two or more rays (i.e. at the spots of increased light intensity — on caustics), Fig. 46.

Nonvalidity of approximation (3.29a) may be perceived in a different way. As already mentioned the main contribution to an oscillation integral of the type (3.24) comes primarily from the points at which $\nabla f = 0$, that is from the critical points of the phase f. According to approximation (3.29a), at such a point $n(\mathbf{x})$ vanishes and equation (3.31b) leads to a divergence.

Hence, equation (3.29a) is not a good approximation of equation (3.28a), even for large k, despite the fact that the neglected term is very small (except for the points at which $a(\mathbf{x})$ vanishes). Approximate equations for $\varphi(\mathbf{x})$, (3.31), fail completely on the caustics, i.e. in spots where the intensity of radiation is highest. The term optical catastrophe has appeared in connection with just this situation. The defect of approximation (3.29) is serious; it cannot be removed by an addition to the right-hand side of equation (3.29b) of corrections proportional to the powers $(1/k)^2$. Equation (3.29b) should thus be replaced by an expression whose analytic dependence on k nearby the caustics would be totally different. As will appear later, the use of a proper approach removes the divergence of the function of radiation intensity on the caustics, the maximum of intensity occurring just on the caustics — at the points satisfying the condition (3.32).

To obtain an analytic expression for the function $\varphi(\mathbf{x})$ behaving properly on the caustics, we shall use the principle of superposition. A function of radiation at the point $\mathbf{c}$, $\Phi(\mathbf{c})$ will be represented as an integral of elementary contributions of the type (3.27)

$$\Phi(\mathbf{c}) = (k/2\pi)^{n/2} \int a(\mathbf{x}, \mathbf{c}) e^{ikf(\mathbf{x};\mathbf{c})} d^n\mathbf{x} \tag{3.33}$$

This is an integral of the form (3.24) and the intensity of radiation at the point $\mathbf{c}$, $I(\mathbf{c})$ is computed according to the equation

$$I(\mathbf{c}) = \Phi^2(\mathbf{c}) \tag{3.34}$$

A normalizing factor in (3.33) was introduced so that the expression for Φ was not divergent for $k \to \infty$; such a form of normalization will be substantiated below.

Let us examine the contribution to the integral (3.33) coming from the vicinity of an arbitrary point $\mathbf{x}_0$. For the purpose, let us represent equation (3.33) in another, equivalent form

$$\Phi(\mathbf{c}) = (k/2\pi)^{n/2} \int e^{(ikf(\mathbf{x};\mathbf{c}) + \ln[a(\mathbf{x};\mathbf{c})])} d^n\mathbf{x} \tag{3.35}$$

and expand the functions f, $\ln(a)$ in the Taylor series near the point $\mathbf{x}_0$, neglecting terms of order higher than first

$$f(\mathbf{x}; \mathbf{c}) \cong f(\mathbf{x}_0; \mathbf{c}) + d\mathbf{x} \cdot \nabla f \qquad (3.36a)$$

$$\ln[a(\mathbf{x}; \mathbf{c})] \cong \ln[a(\mathbf{x}_0; \mathbf{c})] + d\mathbf{x} \cdot \nabla \ln(a) \qquad (3.36b)$$

where $d\mathbf{x} = \mathbf{x} - \mathbf{x}_0$. Eventually, we obtain

$$\Phi(\mathbf{c})_{\mathbf{x}_0} = (k/2\pi)^{n/2} a(\mathbf{x}_0; \mathbf{c}) e^{ikf(\mathbf{x};\mathbf{c})} \int e^{[ik\nabla f + \nabla \ln(a)] \cdot d\mathbf{x}} d^n\mathbf{x} \qquad (3.37)$$

where integration is carried over a certain neighbourhood $N(\mathbf{x}_0)$ of the point $\mathbf{x}_0$. The above integral, for $k \to \infty$ and for the area of the neighbourhood $N(\mathbf{x}_0)$ approaching zero, may be estimated as the product of one-dimensional integrals of the form

$$(k/2\pi)^{1/2} \int_{\delta x(r) - \varepsilon}^{\delta x(r)\varepsilon} \exp\{(ikA + B)\delta x)dx$$

$$= (k/2\pi)^{1/2} \frac{\exp[(ikA + B)\varepsilon] - \exp[-(ikA + B)\varepsilon]}{ikA + B} \xrightarrow[k \to \infty]{} 0 \ (A \neq 0) \ (3.38)$$

where $\delta x = x - x_0$, $\varepsilon \ll 1$, $A = \partial f/\partial x$, $B = \partial[\ln(a)/\partial x$, x is an arbitrary component of the vector $\mathbf{x} = (x_1, ..., x_n)$.

On the other hand, when $A = 0$, i.e. $\partial f/\partial x = 0$ which corresponds to a critical point of a function f, the contribution from the integral (3.38) to the limit $k \to \infty$, is non-zero. In other words, only the critical points $\mathbf{x} = \mathbf{x}^{(r)}$ of the function $f(\mathbf{x}; \mathbf{c})$ contribute to the integral (3.37)

$$\Phi(\mathbf{c}) = \Sigma_r (k/2\pi)^{n/2} \int a[\mathbf{x}^{(r)}; \mathbf{c}] e^{ikf[\mathbf{x}(r) + d\mathbf{x};\mathbf{c}]} d^n\mathbf{x} \qquad (3.39a)$$

$$\nabla \mathbf{x} f[\mathbf{x}^{(r)}; \mathbf{c}] = 0 \qquad (3.39b)$$

Noticeably, the phase f plays the role of a potential function, the requirement $\nabla f = 0$ defines the catastrophe surface M, $\mathbf{c}$ corresponds to control parameters. The radiation intensity at the point $\mathbf{c}$ is represented by the oscillation integral (3.39a), see equation (3.34). The method of estimation of integrals of this type presented above is called the stationary phase method.

Knowing by now that the main contribution to the integral (3.37) is of the form (3.39), we shall try to estimate its magnitude. Let us expand the phase $f(\mathbf{x}; \mathbf{c})$ in the vicinity of the critical point $\mathbf{x}_0$ (i.e. that satisfying the requirement (3.39b) in the Taylor series

$$f(\mathbf{x}; \mathbf{c}) \cong f(\mathbf{x}_0; \mathbf{c}) + 1/2 \, \Sigma_{ij} f_{ij}(\mathbf{x}_0; \mathbf{c}) dx_i dx_j \qquad (3.40)$$

where $f_{ij} \equiv \partial^2 f/\partial x_i \partial x_j$, linear terms at the critical point $\mathbf{x}_0$ being obviously zero (see equation (3.39b)) and terms of third and higher order being neglected. This corresponds to an assumption that the critical point $\mathbf{x}_0$ is nondegenerate. Note that the expansion (3.40) is more accurate than (3.36), since it accounts for higher-order terms.

In addition, we shall assume that the amplitude $a(\mathbf{x}; \mathbf{c})$ is a slow varying function in the vicinity of $\mathbf{x}_0$.

Under these assumptions, we obtain from (3.39a), (3.40):

$$\Phi(\mathbf{c}) \cong (k/2\pi)^{n/2} a(\mathbf{x}_0; \mathbf{c}) e^{ikf(\mathbf{x};\mathbf{c})} \int \exp\left[ik\,1/2\Sigma_{ij} f_{ij}(\mathbf{x}_0; \mathbf{c}) dx_i dx_j\right] d^n x \qquad (3.41)$$

The quadratic form f_{ij} occurring in the sum in (3.41) may be diagonalized by (linear) orthogonal transformation of variables, $x_i \to x_i'$,

$$\Sigma_{ij} f_{ij} dx_i dx_j = \Sigma_i \lambda_i (dx_i')^2 \qquad (3.42)$$

at the same time, since f is the Morse function ($\mathbf{x}_0$ is a nondegenerate critical point of f), its eigenvalues are non-zero, $\lambda_i \neq 0$. On changing variables to x' in the integral (3.41), we obtain (the Jacobian of orthogonal transformation is equal to one)

$$\Phi(\mathbf{c}) \cong (k/2\pi)^{n/2} a(\mathbf{x}_0; \mathbf{c}) \exp\left[ikf(\mathbf{x}_0; \mathbf{c})\right] \int \exp\left[ik\,1/2\Sigma_i \lambda_i (dx_i')^2\right] d^n \mathbf{x}'$$

$$= (k/2\pi)^{n/2} a(\mathbf{x}_0; \mathbf{c}) \exp\left[ikf(\mathbf{x}_0; \mathbf{c})\right] \prod_{j=1}^{n} \int_{-\infty}^{\infty} \exp\left[ik\,1/2\lambda_j (dx_j')^2\right] dx_j'$$

$$\qquad (3.43)$$

There still remains the problem of computation of one-dimensional integrals,

$$\int_{-\infty}^{\infty} \exp(i\lambda x^2) dx = \int_{-\infty}^{\infty} \cos(\lambda x^2) dx + i \int_{-\infty}^{\infty} \sin(\lambda x^2) dx \qquad (3.44)$$

which are called the Fresnel integrals and amount to

$$\int_{-\infty}^{\infty} \exp(\pm i|\lambda|x^2) dx = \sqrt{\pi/2|\lambda|} \pm i\sqrt{\pi/2|\lambda|} = e^{\pm \pi/4}\sqrt{\pi|\lambda|} \qquad (3.45)$$

which finally yields the following result

$$\Phi(\mathbf{c}) \cong a(\mathbf{x}_0; \mathbf{c}) \exp\left[ikf(\mathbf{x}_0; \mathbf{c})\right] \frac{e^{it\Pi/4}}{\displaystyle\prod_{i=1}^{n} |\lambda_i|^{1/2}}$$

$$= a(\mathbf{x}_0; \mathbf{c}) \exp\left[ikf(\mathbf{x}_0; \mathbf{c})\right] \frac{e^{it\Pi/4}}{\left|\det\left[f_{ij}(\mathbf{x}_0; \mathbf{c})\right]\right|^{1/2}} \qquad (3.46)$$

The normalizing factor $(k/2\pi)^{n/2}$ is missing from (3.46) — it vanishes on computing the Fresnel integrals; t stands for the sum of signs of the eigenvalues λ_i (see equation (3.45)), a determinant appears because orthogonal transformation does not change the determinant of the matrix f_{ij}.

It follows from (3.46) that when f has a degenerate critical point at $\mathbf{x}_0$, the denominator vanishes and the functions $\Phi(\mathbf{c})$, $I(\mathbf{c})$ are divergent, that is, such a point $\mathbf{x}_0$ lies on the caustic. In this way, a very close relation between the short-wave optics and elementary catastrophe theory is obtained. More specifically, the caustic is set by the conditions $\nabla f = 0$, $\det|f_{ij}| = 0$, that is it corresponds to the singularity set Σ. When the caustic is observed on a screen at the point $\mathbf{c}$, which corresponds to a projection of the set Σ on the control parameters space, it is the bifurcation set B which is then directly observed on the screen.

The bifurcation set corresponding to a cusp catastrophe (A_3) may be observed in the case of diffraction of rays reflected from an internal surface of, for example, a cup illuminated diagonaly from above. Changes in the form of the caustic showing in the bottom of the cup can be observed on varying the angle of inclination of the cup with respect to the light source (these are the caustics corresponding to the bifurcation set of a more complex elementary catastrophe), see Figs. 7, 8.

Equation (3.46) fails on the caustics, resulting in the infinite light intensity. This is due to restricting the Taylor expansion (3.40) to quadratic terms. As follows from Chapter 2, in the case when a function has a degenerate critical point, higher-order terms should be accounted for in the Taylor expansion. Hence, to estimate the function $\Phi(\mathbf{c})$ on the caustic, instead of the Taylor expansion (3.40) or (3.36), a more accurate expansion has to be used. In the vicinity of the degenerate critical point $\mathbf{x}_0$ we expand f in the Taylor series

$$f(\mathbf{x}; \mathbf{c}) \cong f(\mathbf{x}_0; \mathbf{c}) + F(\mathbf{x}_0; \mathbf{c}) \qquad (3.47)$$

from which linear and quadratic terms are absent, and F corresponds to one of the elementary catastrophe functions. Substitution of the expansion (3.47) into (3.39a) yields

$$\Phi(\mathbf{c}) \cong (k/2\pi)^{n/2} a(\mathbf{x}_0) \exp[ikf(\mathbf{x}_0; \mathbf{c})] \int \exp[ikF(\mathbf{x}; \mathbf{c})] d^n\mathbf{x} \qquad (3.48)$$

The methods of computing integrals of the type (3.48), where F corresponds to one the potential functions of the Thom elementary catastrophe, will be discussed in Section 4.4.4. Note also that if the observed caustic is structurally stable, it should be expressed by equation (3.48).

3.4.3 The short-wave limit for the Schrödinger equation

Quantum phenomena of physics and chemistry may be described by the Schrödinger equation, which for a particle of mass m is of the form

$$-\hbar^2/2m\nabla^2\Psi(\mathbf{x}) + V(\mathbf{x})\Psi(\mathbf{x}) = E\Psi(\mathbf{x}) \qquad (3.49)$$

where $\hbar$ is Planck's constant divided by 2π, $V(\mathbf{x})$ is the potential energy and E is the energy of state Ψ. For simplicity, the short-wave approximation will be considered for a particle moving in one dimension only

$$-\hbar^2/2m \, d^2\Psi/dx^2 + V(x)\Psi(x) = E\Psi(x) \qquad (3.50)$$

We shall look for solutions to equation (3.50) having the form analogous with (3.27)

$$\Psi = \exp(i/\hbar S) \qquad (3.51)$$

Putting in (3.51) into the Schrödinger equation (3.50) yields

$$1/2m(dS/dx)^2 - i\hbar/2m \, d^2S/dx^2 = E - V(x) \qquad (3.52)$$

whose solution will be expressed as a series in a small parameter $\hbar/i$

$$S = S_0 + \hbar/i S_1 + 1/2(\hbar/i)^2 S_2 + \dots \qquad (3.53)$$

Substitution of series (3.53) into equation (3.52) and collection of terms of the same order in $\hbar/i$ yields first- and second-order equations

$$1/2m(dS_0/dx)^2 = E - V(x) \qquad (3.54)$$

$$(dS_1/dx)(dS_0) + 1/2(d^2S_0/dx^2) = 0 \qquad (3.55)$$

and equations of higher order.

By the relationship between the energy E and momentum $p(x)$

$$E = V(x) + p^2(x) \tag{3.56}$$

equation (3.54) becomes

$$dS_0/dx = \pm p(x), \quad p(x) = \sqrt{2m[E - V(x)]} \tag{3.57}$$

The solution of equation (3.57) can be obtained by elementary integration

$$S_0 = \pm \int_{x_0}^{x} p(y)dy \tag{3.58}$$

After determining S_0, S_1 can be calculated from equation (3.55)

$$(dS_1/dx)[\pm p(x)] + 1/2[\pm dp(x)/dx] = 0 \tag{3.59}$$

obtaining, after carrying out elementary integration, the equation for S_1

$$S_1(x) = -1/2\ln|p(x)| + C \tag{3.60}$$

where C is the integration constant.

Hence, the wave function Ψ may be approximately expressed by the equation analogous with (3.46)

$$\Psi(x) = C_1 \exp\left[i/\hbar \int_{x_0}^{x} p(y)dy - 1/2\ln|p(x)|\right] +$$

$$+ C_2 \exp\left[-i/\hbar \int_{x_0}^{x} p(y)dy - 1/2\ln|p(x)|\right]$$

$$= \frac{C_1}{\sqrt{p(x)}} \exp\left[i/\hbar \int_{x_0}^{x} p(y)dy\right] + \frac{C_2}{\sqrt{p(x)}} \exp\left[-i/\hbar \int_{x_0}^{x} p(y)dy\right] \tag{3.61}$$

Near classical points of return, defined for a given potential $V(x)$ by the equation

$$p(x) = \sqrt{2m[E - V(x)]} = 0 \tag{3.62}$$

i.e. $V(x) = E$, equation (3.61) cannot be applied. Classical points of return are thus the analogues of caustics for light rays. Near the return points a better approximation for the wave function is needed than that deriving from the method described above.

A closer inspection of the shape of Ψ in the vicinity of the return point x_0, i.e. that satisfying equation (3.62), will now be performed. Expand the potential $V(x)$ as a series in powers of $x - x_0$

$$V(x) \cong E + (x - x_0)\mathrm{d}V/\mathrm{d}x(x_0) \qquad (3.63)$$

so that the Schrödinger equation nearby the return point x_0 has an approximate from

$$\mathrm{d}^2\Psi/\mathrm{d}x^2 - 2m/\hbar^2 |F|(x - x_0)\Psi = 0 \qquad (3.64)$$

where $F = (\mathrm{d}V/\mathrm{d}x)(x_0)$. Changing variables

$$x \to u = (x - x_0)(2m|F|/\hbar^2)^{1/3} \qquad (3.65)$$

we reduce equation (3.64) to the form

$$\mathrm{d}^2\Psi/\mathrm{d}u^2 - u\Psi = 0 \qquad (3.66)$$

called the Airy equation (an identical equation appears at the description of light diffraction on the edge of an obstacle).

The solution to the Airy equation may be written in terms of the oscillatory integral (see Section 3.4.5)

$$\Psi(u; k) = (k/2\pi)^{1/2} \int\limits_{-\infty}^{\infty} \exp\left[ik(1/3\, x^3 + ux)\right] \mathrm{d}x = 2\left[k/(2\pi)^3\right]^{1/6} \mathrm{Ai}(k^{2/3}u)$$

$$(3.67)$$

where Ai is a special function defined by the last equality in equation (3.67).

Ultimately, the approximate Schrödinger equation (3.64), exact near the return point x_0, has a solution which can be written in terms of the diffraction integral of the potential function $F(x; u) = 1/3\, x^3 + ux$ of a fold catastrophe (A_2), see equations (3.67), (3.25).

As follows from our previous considerations, the integral (3.67) cannot be computed by the stationary phase method in the vicinity of the caustic due to appearance of the divergence associated with having by the potential function $F(x; u)$, i.e. the phase, a degenerate critical point on the caustic (at the return point x_0). However, the Airy function can be computed by another method, see Section 3.4.5. The form of the Airy function is shown in Fig. 47. Let us recall that $\Psi^2(u; k)$ is interpreted as the intensity of scattered light or the probability of finding the particle at the point u.

In the shape of Ai(y), three zones can be distinguished. In the case of light diffraction on an obstacle, region I, in which the function Ai(y) oscillates, corresponds to an illuminated zone; region II, wherein Ai(y)

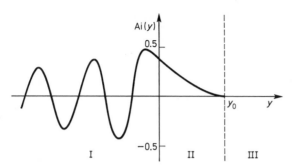

Fig. 47. The Airy function.

decreases exponentially, corresponds to a shadow zone, while region III corresponds to a nonilluminated zone. Interestingly, the maximum of light intensity is before the obstacle and, consequently, the light intensity begins to decrease already before the obstacle.

In the case of a quantum potential barrier, zone I corresponds to the solution inside the potential well, zone II represents an exponential decay of the wave function inside the potential barrier, in zone III the wave function is practically zero. For the potential barrier of a width smaller than y_0, see Fig, 47, the effect of „tunneling" through the barrier is possible.

3.4.4 Examples of higher-order diffraction catastrophes

Diffraction integrals are an essential tool of the description of a variety of diffraction phenomena in opitcs and quantum mechanics. The intensity of scattered light or the probability of finding a particle may be represented by integrals of the form (3.25). Let us recall that $|\Phi|^2$ can be interpreted as the light intensity or the density of probability of finding a particle.

For example, the diffraction integral (3.25) containing the potential function of a cusp catastrophe (A_3), $F(x; \mathbf{c}) = x^4 + ax^2 + bx$, describes ligth scattering on a two-dimensional diffraction grating, see Fig. 48. A function defined by such an integral is called the Pearcey function.

The above pattern of intensity of scattered light may be observed by viewing a strong point light source placed behind a properly matted glass panel (having a rectangular pattern of grooves on the surface, Fig. 48a).

There exist many other phenomena described by diffraction integrals. For example, in optics the description of light scattering on water droplets

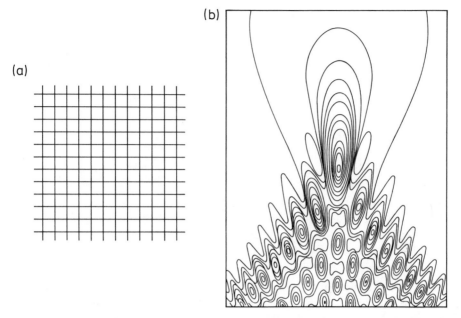

Fig. 48. Diffraction cusp catastrophe (A_3): (a) diffraction grating; (b) oscillation integral A_3.

or in quantum chemistry the description of some type of collisions of molecules lead to the canonical integral (3.25) containing the potential function $F(\mathbf{x}; \mathbf{c})$ of the elementary catastrophe of elliptic umbilic (D_4^{-}): $F(\mathbf{x}; \mathbf{c}) = x^3 - 3xy^2 + a(x^2 + y^2) + bx + cy$ (Fig. 49).

Fig. 49. Diffraction elliptic umbilic catastrophe (D_4^{-}): diffraction integral D_4^{-} in the $b, c, a = 0$ plane.

On the other hand, in connection with theoretical analysis of the problems related to scattering of atoms on crystals and in the model of scattering of atoms on rigid rotators appears the canonical integral (3.25) containing the potential function $F(\mathbf{x}; \mathbf{c})$ of the elementary catastrophe of hyperbolic umbilic $(D_4{}^+)$: $F(\mathbf{x}; \mathbf{c}) = x^3 + y^3 + axy + bx + cy$.

3.4.5 Computation of oscillatory integrals

Let us recall that in the case of computing the oscillatory integrals (3.25) in which the function F has a degenerate critical point, the stationary phase method, described in Section 3.4.2, fails. There are two basic methods of computation of such integrals which will be exemplified by the Airy function. The information provided below is brief and intended to facilitate the reader an access to suitable references (see bibliographical remarks at the end of this chapter).

The first method involves finding a differential equation satisfied by a function given in the form of an oscillatory integral solving directly this equation. The Airy function defined by equation (3.67) may be shown to fulfil the differential equation (3.66). There are many effective methods of solving differential equations numerically. Thus, finding the equation satisfied by a given diffraction integral enables its effective computation.

The second method consists in writing an oscillatory integral in the form

$$\Psi(\alpha) = \oint_c f(z) e^{\alpha h(z)} \, dz \qquad (3.68)$$

where α is a large positive real number, C is the integration contour in the conjugate plane z, $f(z)$ and $h(z)$ are analytic functions of z. The integrand in (3.67) has a suitable form, $f(z) = (k/2\Pi)^{1/2}$, $\alpha h(z) = ik(z^3/3 + uz)$ and it can be demonstrated that there exists such a contour C that equation (3.68) is reduced to equation (3.67). An integral of the form (3.68) may be estimated by the so-called saddle node method. The idea underlying the method is as follows. In the case of integral of a real variable of the form

$$f(\alpha) = \int_a^b f(x) e^{\alpha h(x)} \, dx \qquad (3.69)$$

where $h(x)$ has a maximum in the integration interval, for example at the point $x_0 \in [a, b]$ and $\alpha \gg 1$, the main contribution to the integral comes primarily from the neighbourhood of x_0. Likewise, in the case of integral (3.68) for $\alpha \gg 1$ the main contribution derives from this part of the contour C on which the real part of $h(z)$ is large compared with other fragments of the contour. It follows from the theory of conjugate functions that the path of integration, over which the real part of the analytic function $h(z)$ changes most rapidly, passes through the saddle node satisfying the requirement

$$\mathrm{d}h(z)/\mathrm{d}z = 0 \qquad\qquad (3.70)$$

3.5 THE ZEEMAN MODELS FOR THE HEARTBEAT AND THE NERVE IMPULSE TRANSMISSION

Investigations of the heartbeat have revealed that the heart may occur in two fundamental states: the state of decontraction (diastole) and the state of contraction (systole). Responding to an electrochemical stimulation, each fibre of the cardiac muscle rapidly contracts, remaining in this state momentarily, followed by a rapid return to the state of decontraction.

In the case of transmission of nerve impulses, the dynamics is different. The state of a transmitting nerve fibre, axon, is determined by the electrochemical potential between inner and outer fibres of the axon. In the absence of a perturbation the potential remains at a constant level. In the case of impulse transmission, the potential abruptly changes, followed by a slow a return to the initial state.

Zeeman's concept of modelling these phenomena involves their qualitative description by differential equations: phase portraits of these equations must only meet some qualitative requirements consistent with the above characteristic of the systems being modelled. The processes described above have certain significant characteristics, which have to be taken into account in the model:

(a) there is stable stationary state to which the system returns periodically,

(b) there exists a mechanism of loss of stability of the stationary state,

(c) there exists a mechanism of return to the stationary state.

Note that the described processes differ only in (c). The requirements which have to be satisfied by the phase portrait of a system of differential equations describing such processes should now be examined (it should be pointed out that the Zeeman method of modelling is well suited for modelling equations of chemical kinetics).

Let us now consider, after a Zeeman paper, the following system of equations (a dot denotes the derivative with respect to time)

$$\dot{x} = -\lambda x \tag{3.71a}$$

$$\dot{b} = -b \tag{3.71b}$$

where $\lambda \gg 1$. The phase trajectories of system (3.71) are illustrated in Fig. 50.

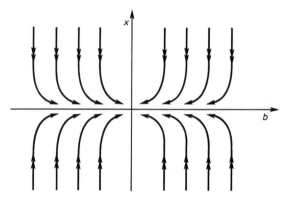

Fig. 50. Phase portrait for the system (3.71).

Equation (3.71a) describes a rapid evolution $x(t) \rightarrow x = 0$: all trajectories, except for those lying along the b-axis, are almost parallel to the x-axis and approach the b-axis, whereupon becoming parallel to the slow motion

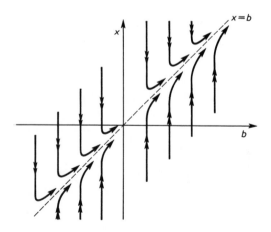

Fig. 51. Phase portrait for the system (3.72).

described by equation (3.71b). The point $x = 0$ is called a slow manifold. The system (3.71) has the stationary solution: $\dot{x} = 0$, $\dot{b} = 0$ – a singular point of the coordinates (0, 0).

A similar phase portrait has the system

$$\varepsilon\dot{x} = -(x + b) \tag{3.72a}$$

$$\dot{b} = x \tag{3.72b}$$

where $\varepsilon = \lambda^{-1} \ll 1$. A slow manifold is given by the equation $x + b = 0$; this set is approached by the trajectories of the system by way of rapid processes described by equation (3.72a), see Fig. 51.

The singularity of the system (0, 0), is an unstable stationary point for $\varepsilon > 0$. Apparently, equations (3.71), (3.72) can be immediately solved, being the system of homogeneous first-order linear equations. The solutions have the form

$$x(t) = x_0 e^{-\lambda t}, \quad b(t) = b_0 e^{-t} \tag{3.71c}$$

$$b(t) = c_1 e^{r_1 t} + c_2 e^{r_2 t}, \quad x(t) = r_1 c_1 e^{r_1 t} + r_2 c_2 e^{r_2 t}$$

$$r_1 \cong -1, \quad r_2 \cong -(1/\varepsilon) \quad (0 < \varepsilon \ll 1) \tag{3.72c}$$

After a short time the solutions, for $\varepsilon > 0$, approach

$$x(t) \cong 0, \quad b(t) \cong b_0 e^{-t} \tag{3.71d}$$

$$x(t) \cong -c_1 e^{-t}, \quad b(t) \cong c_1 e^{-t}, \quad x(t) + b(t) \cong 0 \tag{3.72d}$$

A modification of the system (3.72a), (3.72b) having the form

$$\varepsilon\dot{x} = -(x^3 - x + b) \tag{3.73a}$$

$$\dot{b} = x \tag{3.73b}$$

$\varepsilon \ll 1$, aims at transforming the system (3.73a), (3.73b), having trajectories escaping to infinity, to the form wherein the phase trajectories will remain in a certain region of finite dimensions. That is the case for the system (3.73), since the term $-x^3$, negligible for small x, predominates for $x \gg 1$, which leads to a decrease in the x value. The form of a slow manifold results from the condition $\varepsilon\dot{x} = 0$ or

$$x^3 - x + b = 0 \tag{3.74}$$

Knowledge of the slow set enables qualitative understanding of a phase portrait for the system of equation (3.73), see Fig. 52.

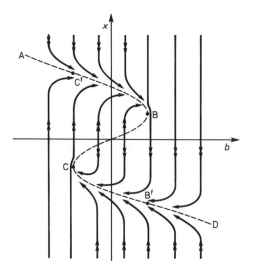

Fig. 52. Slow surface for the system (3.73).

Apparently, the fragments AB and CD of the slow set are attracting (attractor), whereas the fragment BC is repelling (repeller). A point being in an arbitrary spot of the phase space rapidly shifts towards C′B, B′C, where it becomes confined; the evolution from C′ to B proceeds slowly, from B to B′

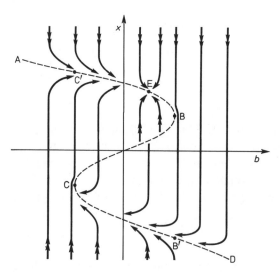

Fig. 53. Phase portrait for the system (3.75).

rapidly, from B′ to C slowly and from C to C′ rapidly. The trajectory C′BB′C is a so-called limit cycle.

Zeeman carried out a further modification of the system (3.73)

$$\varepsilon\dot{x} = -\left(x^3 - x + b\right) \tag{3.75a}$$

$$\dot{b} = x - x_0 \tag{3.75b}$$

where $x_0 > 1/\sqrt{3}$, to model the heartbeat more realistically. A phase portrait for the system (3.75) is shown in Fig. 53.

At the points B, C the variable x has a value $\pm\sqrt{3}$. The system (3.75) has one stationary point $E = \left(x_0, b_0\right)$, $b_0 = x_0 - x_0^3$, whose coordinates, result from the conditions $\dot{x} = 0$, $\dot{b} = 0$. Linearization of the system (3.75) in the vicinity of E, $x \cong x_0 + \xi$, $b = b_0 + \beta$, $|\xi|$, $|\beta| \ll 1$,

$$\varepsilon\dot{\xi} \cong \left(x_0 + \xi\right) - \left(x_0 + \xi\right)^3 - \left(b_0 + \beta\right) \cong \left(1 - 3x_0^2\right)\xi - \beta \tag{3.76a}$$

$$\dot{\beta} \cong \left(x_0 + \xi\right) - x_0 = \xi \tag{3.76b}$$

leads to a condition that the point $E = \left(x_0, x_0 - x_0^3\right)$ is a stable stationary point for $1 - 3x_0^2 < 0$.

Then x variable plays in Zeeman's model the role of length of a fibre of the cardiac muscle while the b variable corresponds to the electrochemical control (contraction of the cardiac muscle is triggered by a biochemically generated electric impulse). A stable stationary point E may occur near the point B which is infinitely sensitive to perturbations. To transfer the system from the stable stationary point E to B, a perturbation of the system is required; if E is located close to B the perturbation can be small. The mechanism of switching the heart from the state of equilibrium E (lack of heartbeat) to the state of action involves removing the system from the state E to B by way of stimulation, for example by an electric impulse. On reaching the state B the model system imitates the heartbeat — this is the trajectory BB′CC′E. A subsequent cycle requires the repeated stimulation at the point E.

We will now introduce to equations (3.76) the second control parameter, a, standing for tension (tone) of the cardiac muscle and constant for a given state of the heartbeat dynamics:

$$\varepsilon\dot{x} \doteq -\left(x^3 + ax + b\right) \tag{3.77a}$$

$$\dot{b} = x - x_0 \tag{3.77b}$$

$$\dot{a} = 0 \tag{3.77c}$$

In this way, the state of the system (3.77) occurs, in the (x, a, b) space, on the slow manifold given by the equation $\varepsilon\dot{x} = 0$, that is

$$x^3 + ax + b = 0 \tag{3.78}$$

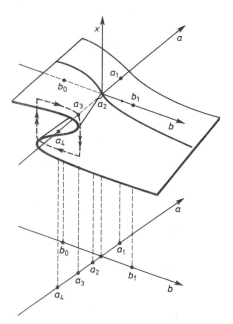

Fig. 54. Slow surface for the heartbeat model.

The slow surface (3.78) corresponds to the catastrophe surface M_3 of a cusp catastrophe (A_3), see Fig. 54 in which the values of the control parameter which cannot be exceeded in a living organism are marked on the b-axis (recall that b is related to a biochemical state of the heart muscle). The four fundamental states of dynamics of the heartbeat may now be described in terms of the value of parameter a (tension of the cardiac muscle). Since the parameter a is constant for a given state, the states of the system lie on the sections $a = \text{const}$ of the cusp catastrophe surface. The four states are conveniently represented in the (x, b) plane, Fig. 55, the parameter x_a having to have such a value that the linearized system (3.76) does not have a stable stationary point. The four states shown in Fig. 55 have the following interpretation:

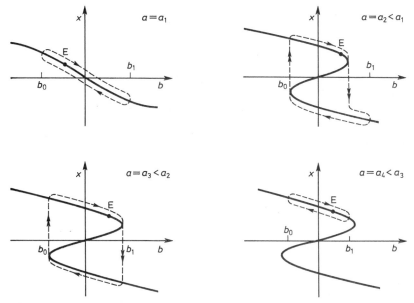

Fig. 55. Dynamical states of the heartbeat.

(1) $a = a_1 > 0$ — the heartbeat is impossible, the heart makes smooth movements. Such a state is called ventricular fibrillation; it may occur at a large loss of blood and a decreased blood pressure;

(2) $a = a_2 < a_1$ — the heartbeat of a small amplitude (not the whole available variability range of the parametr b is used);

(3) $a = a_3 < a_2$ — the heartbeat of a large amplitude, employing an entire variability range of the electrochemical control b (ventricular beating);

(4) $a = a_4 < a_3$ — ventricular flutter: the state of a permanent contraction (the heart in this state makes movements of a small amplitude without beating).

Such a state may take place after administering a large dose of caffeine (excessive stimulation of tension of the cardiac muscle). A variability range of the parameter b is too small for the heartbeat to occur.

Let us now proceed to modelling of the process of impulse transmission by the nerve or, more specifically, by the axon — a long fibre joining the receptor to the nerve cell. Hodgkin and Huxley have carried out studies on nerve impulse transmission by the exceptionally long optical nerve of a squid. Zeeman has succeeded in modelling an action of the axon on the

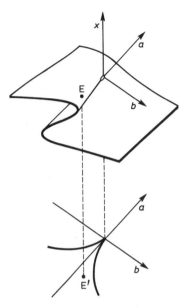

Fig. 56. Slow surface for the model of nerve impulse transmission by the axon.

cusp catastrophe (A_3) surface. The required characteristics of the model were
described at the beginning of the subchapter (items a, b, c, p.118); a salient
feature of the considered dynamics is a slow return to the equilibrium state
(after a rapid departure from the stationary state). In the Zeeman model the
x variable corresponds to the Na^+ ion conductance across the axon
membrane, a is the K^+ ion conductance while b represents the membrane
potential, see Fig. 56.

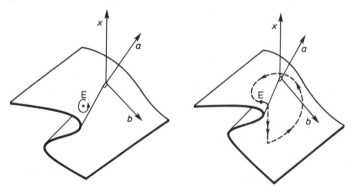

Fig. 57. Dynamical states of the axon on the surface of the cusp catastrophe.

Zeeman has given a system of differential equations, describing time evolution of the variables x, a, b, for which a slow surface is the cusp catastrophe surface (3.78). The system of Zeeman equations has an attracting and stable stationary point E. The process of nerve impulse transmission represented on the catastrophe surface M_3 is shown in Fig. 57.

At the first stage, stimulation of the stable state E by an external impulse acting on the receptor takes place. If the stimulation value (generating a change in the parameter b) is too small, the system returns to the point E. If the stimulation value exceeds a critical value, the system evolves by falling rapidly on the lower lobe of the catastrophe surface M_3, followed by a slow evolution to the attracting state E. On returning to the state E, a repeated stimulation of the axon is possible. Apparently, the system has a threshold stimulation value (perturbation of the membrane potential b). On exceeding this value, the sodium ion conductance across the membrane x abruptly varies.

On the basis of the model constructed in this way Zeeman calculated the speed of impulse transmission by the axon, obtaining a considerably better agreement with the experiment than in the original paper by Hodgkin and Huxley.

3.6 POPULATION MODELS AND SEQUENTIAL MODELS

3.6.1 Introduction

Sequential models are represented by recurrent equations of the form

$$x_{n+1} = F(x_n, y_n, ...; a, b, ...) \tag{3.79a}$$

$$y_{n+1} = G(x_n, y_n, ...; a, b, ...) \tag{3.79b}$$

where the state variables x, y, ... are computed in the step $n + 1$ on the basis of the values of these variables obtained in nth step (and, possibly, on the basis of the values of the variables from the step $n - 1$, etc.) and the known initial values from equations of the form (3.79). The functions F, G, ... are often non-linear and depend on the control parameters a, b, ...

Consider a sequential model dependent only on one state variable x. The sequence x_1, x_2, x_3, ..., defined for specific values of control parameters and for a given initial value x_1 will be called a (sequential) process:

$$x_{n+1} = F(x_n; a, b, ..., h, ...) \tag{3.80}$$

The nature of a process generally depends on the values of control parameters and may vary in a discontinuous manner with continuous changes in parameters — such a phenomenon will be referred to as a catastrophe. It follows from the notation of equation (3.80) in the form

$$(x_{n+1} - x_n)/h = [F(x_n) - x_n]/h \equiv dG(x_n)/dx_n, \qquad 0 < h < 1 \tag{3.81a}$$

that the function G may be related to the potential function V in the equation

$$dx/dt = dV/dx, \quad \text{i.e.} \ \ dV/dx = [F(x_n) - x_n]/h \tag{3.81b}$$

Sequential models comprise, for example, ecological models describing the evolution with time of a certain population. For instance, the process of development of a periodically reproducing population can be represented by the model (3.80). There are numerous other dependences of this type in biology, e.g. in genetics, where equation (3.80) (or equations (3.79)) describes changes in the occurrence rate of a gene, in epidemiology where x_n is a fraction of the nth population infected with a disease or a fraction of a population in the nth month of spread of an epidemic, etc. Examples deriving from economics (e.g. in theory of economic cycles) or sociology (e.g. in theory of learning or spread of a gossip) are also known.

In Chapter 2 we have shown that the equation of type (3.80) may appear in models describing the dynamics of a chemical reaction (see also remarks in Section 3.7). The model (3.80) will be employed in Chapter 6 for the description of changes in the dynamics of the Belousov–Zhabotinskii reaction.

We shall now continue with the examination of a simple model of type (3.80), which has many applications. In the present section, an ecological interpretation will be described. By means of equation (3.80) the process of growth of a periodically reproducing population can be described provided that its successive generations, $x_1, x_2, ...$, do not overlap. A behaviour of this type can take place in the case of insects, whose consecutive generations hatch in spring from eggs laid in fall, mature individuals not surviving winter. In a number of biological models the dependence of the function F on x is as follows: when x is small, $F(x)$ is an increasing function whereas when x is large, $F(x)$ is a decreasing function. Such a behaviour, for example for the population of insects, may be explained in the following way. When the abundance of nth population is small (small x_n), there exist propitious conditions for is growth (sufficient quantity of food, etc.). By contrast, when the abundance of a population approaches the limit of capacity of the environment, the conditions become unfavourable (lack of food, development of predators, etc.). One of the simplest functions fulfilling the above conditions is the so-called logistic function $F(x; a, b) = x(b - ax)$, where a, b are constants. Then, equation (3.80) takes the form of a logistic difference equation in one state variable X

$$X_{n+1} = F(X_n) = X_n(b - aX_n) \tag{3.82}$$

It is convenient to switch in (3.82) to a normalized state variable $x_n = (a/4b)X_n, x_n \in [0, 1]$,

$$x_{n+1} = 4bx_n(1 - x_n) \tag{3.83}$$

where the constant b must satisfy the requirement $b \in [0, 1]$ so that $x_n \in [0, 1]$. The x_n value may be interpreted as a percent population of the environment with respect to its maximum capacity equal to a/b.

In further analysis we shall concentrate on the examination of equation (3.83). It is worth noting that there are other, more realistic models, in which $F(x)$ has a more elaborate form, for example:

$$x_{n+1} = x_n \exp[r(1 - x_n)] \tag{3.84}$$

Feigenbaum has demonstrated that all sequential models of the type (3.80), in which the function $F(x)$ has a single quadratic maximum, are qualitatively identical to the model (3.83).

3.6.2 Analysis of a logistic sequential process

In equation (3.83) x is a state variable whereas b is a control parametr. As will turn out, the iterative process (3.83) is virtually independent of the initial value x_1. First, let us examine fixed points of the process. Substitution of the stationary state requirement $x_{n+1} = x_n \equiv x$ into (3.83) yields the equation for fixed points

$$x_* = F(x_*), \quad \text{or} \quad x_* = 4bx_*(1 - x_*) \tag{3.85}$$

which has two solutions: $x_*^{(1)} = 0$, $x_*^{(2)} = 1 - (1/4b)$.

Properties of the process (3.83) for small x_n values are easy to find. For sufficiently small x_n, the quadratic term in (3.83) may be neglected and an approximate linear equation

$$x_{n+1} \cong 4bx_n \tag{3.86}$$

is obtained, the solution to which is expressed by

$$x_n \cong (4b)^{n-1} x_1 \tag{3.87}$$

Since the sequence $\{x_n\}$ must be bounded, the parameter b has to satisfy the condition $0 < 4b < 1$. Hence, when $b \in [0, 1/4]$, the sequence $\{x_n\}$ approaches zero, i.e. the attracting fixed point $x_*^{(1)}$ irrespective of the x_1 value (x_1 must be different from $x_*^{(2)}$).

When a continuous change in the control parametr b results in exceeding the value $b_0 = 1/4$, we have a loss of stability by the fixed point $x_*^{(1)}$. The new stable fixed point, $x_*^{(2)} = 1 - (1/4b)$, close to $x_*^{(1)}$ for b close to b_0, appears. Such a catastrophe is called bifurcation. Catastrophic behaviour of the process (3.83) for $b > b_0$ is revealed in the fact that the solution (3.87) for $b = b_0 + \varepsilon$ diverges to infinity for an arbitrarily small positive ε.

To show that for $b > b_0$ the fixed point $x_*^{(2)}$ acquires stability, equation (3.83) will be linearized in the vicinity of this point (approximation of equation (3.83) by (3.86) is linearization in the neighbourhood of the point $x_*^{(1)} = 0$). Hence, let us represent x_n in the form

$$x_n = x_*^{(2)} + y_n = 1 - (1/4b) + y_n \tag{3.88}$$

and substitute into equation (3.83). For a sufficiently small y_n the term proportional to y_n^2 may be ignored, and the following approximate linear equation in the y variable is obtained:

$$y_{n+1} \cong (2 - 4b) y_n \tag{3.89}$$

Hence, the solution to equation (3.83) is given by

$$x_n \cong [1 - (1/4b)] + y_n = [1 - (1/4b)] + (2 - 4b)^{n-1} y_1 \tag{3.90}$$

The sequence $\{y_n\} = (2 - 4b)^{n-1} y_1$ is bounded if $|2 - 4b| < 1$. If this condition is fulfilled, the sequence $\{y_n\}$ approaches zero (monotonically for $(2 - 4b) > 0$ or oscillating for $(2 - 4b) < 0$) and then the sequence $\{x_n\}$ approaches $x_*^{(2)} = 1 - (1/4b)$.

The requirement for boundedness of the sequence $\{y_n\}$, $|2 - 4b| < 1$, is satisfied for $b \in (1/4, 3/4)$. When b exceeds the value $b_1 = 3/4$, the next catastrophe occurs — the fixed point $x_*^{(2)}$ loses stability and the character of the process (3.83) changes qualitatively. It turns out that for $b > b_1$ two attracting fixed points, $x_*^{(3)}$, $x_*^{(4)}$ appear and the sequence $\{x_n\}$ oscillates between them. Such solutions are the simplest generalization of the requirement of a stationary character for the sequence $\{x_n\}$ $(x_{n+1} = x_n)$ and meet the condition $x_{n+2} = x_n \equiv x$. Substitution of this condition into equation (3.83) yields

$$x = F[F(x)], \quad \text{or} \quad x = 4b4bx[1 - 4bx(1 - x)] \tag{3.91}$$

The solutions to equation (3.91) are obviously those solutions to equation (3.85) for the fixed points $x_*^{(1)} = 0$, $x_*^{(2)} = 1 - (1/4b)$ and $x_*^{(3,4)} = = 1/2\{1 + 1/4b[1 \pm \sqrt{(4b + 1)(4b - 3)}]\}$. The solutions $x_*^{(3)}$, $x_*^{(4)}$ are fixed points of the process

$$x_{n+2} = F[F(x_n)], \quad \text{or} \quad x_{n+2} = 4b4bx_n[1 - 4bx_n(1 - x_n)] \tag{3.92}$$

Equation (3.92) can be linearized in the vicinity of $x_*^{(3)}$ or $x_*^{(4)}$, solved and the stability of its solutions examined exactly in the same way as in the case of equation (3.83). The sequence $\{x_n, x_{n+2}, x_{n+4}, ...\}$ appears to approach one of the fixed points $x_*^{(3)}$, $x_*^{(4)}$ for $b \in (b_1, b_2)$, $b_1 = 3/4$, $b_2 = (1 + \sqrt{6}/4$ and, hence, the sequence $\{x_n\} = x_n, x_{n+1}, x_{n+2}, ...$ oscillates between the two attracting fixed points $x_*^{(3,4)}$. Such a behaviour of a sequence is called 2-cycle. The process of approaching a fixed point (e.g. $x_*^{(1)}$ or $x_*^{(2)}$ is referred to as 1-cycle (or 2^0-cycle)).

For $b > b_2$ the fixed points of the 2-cycle lose stability; on the other hand, the fixed points of the process $x_{n+4} = F(F[F\{F(x_n)\}])$ become stable (attracting). Repeating the above analysis we arrive at a conclusion that for $b \in (b_2, b_3)$, for a certain $b_3 < 1$, the sequence $\{x_n\}$ behaves as the 4-cycle (i.e. 2^2-cycle): the sequence has four attracting points and the successive terms of the sequence: $x_n, x_{n+1}, x_{n+2}, x_{n+3}$, approach the respective attracting poins $x_*^{(5)}$, $x_*^{(6)}$, $x_*^{(7)}$, $x_*^{(8)}$.

For $b > b_3$, $b > b_4$, etc., successive losses of stability of the process (3.83) take place; the sequence $\{x_n\}$ has 2^3, 2^4 attracting points. The described phenomenon is called period doubling. Finally, for the value $b_\infty = = 0.892486...$ the process has the nature of the 2^∞-cycle. Feigenbaum has shown that the successive values of the parameter b, for which a qualitative change in nature of the process occurs, satisfy the dependence (δ is called the Feigenbaum constant):

$$\frac{b_{m+1} + b_m}{b_{m+2} - b_{m+1}} \to \delta, \quad \delta = 4.6692... \tag{3.93}$$

Feigenbaum also demonstrated that the process of type (3.80) behaves analogously with the logistic process (3.83) if the function $F(x)$ has in its variability range, e.g. the section $[0, 1]$, juest one quadratic maximum and does not have other critical points.

For $b > b_\infty$ the process becomes nonperiodic except for narrow regions of the parameter b in which the 3-cycles appear. Eventually, for $b = 1$ the process has the character of white noise. In this case an exact solution to the process (3.83) may be given

$$x_n = \sin(2^n a) \tag{3.94a}$$

$$x_1 = \sin^2(a), \quad a = \arcsin\left(\sqrt{x_1}\right) \tag{3.94b}$$

The periodicity requirement for solutions (3.94), $x_{n+k} = x_n$, leads to solutions in the nature of unstable k-cycles

$$a_\pm = N\pi/(2^k \pm 1), \ k = 1, 2, ..., \text{ and } k = 0 \text{ for } a_+, \ N = 0, \ \pm 1, \ \pm 2, ... \tag{3.95}$$

For other values of a the process is nonperiodic and chaotic.

3.7 RELATION OF THE DISCUSSED MODELS TO CHEMICAL SYSTEMS

In Section 1.3 we described the systems in which qualitative and discontinuous changes of state, that is catastrophes, could be observed at a continuous variation in control parameters. The catastrophes occurring in some systems were discussed in terms of elementary catastrophe theory in Sections 3.2–3.6. The discussion was confined to non-chemical systems; such a classification (as we shall see later) being rather artificial. Catastrophes (static and dynamic) occurring in chemical systems will be described in Chapters 5, 6.

Many of the considered problems, such as the problem of stability of soap films, the liquid–vapour phase transition, the diffraction phenomena, descriptions of the heartbeat or the nerve impulse transmission, catastrophes described by non-linear recurrent equations have a close relation to chemical problems.

The stability of thin films and the catastrophes of film systems may play a crucial role in the case of chemical reactions proceeding at the boundary of a liquid phase and another phase. Phase transitions are of a great significance in physical chemistry. The diffraction phenomena for the

Schrödinger equation and the corresponding diffraction catastrophes occurring at collisions of molecules are important from the viewpoint of the description of chemical reactions taking place upon contact (collision) of molecules. As shown in Section 1.3, recurrent equations appear from a description of the kinetics of chemical reactions.

The method used by Zeeman to model the heartbeat and the process of nerve impulse transmission by means of differential equations (see Section 3.5) is frequently employed in chemical kinetics (where it is called the Tikhonov method or the method of stationary concentrations). Application of the method to chemical kinetics equations will be discussed in Chapter 4.

In Chapters 5, 6 we will deal with the relation of catastrophes of a diffraction type for the Schrödinger equation (see Section 3.4) and catastrophes occurring in non-linear sequential systems (see Section 3.6) to the catastrophes taking place in chemical systems.

Bibliographical Remarks

Information on the topology and catastrophes of soap films can be found in Poston's lecture and in papers by Fomyenko.

Phase transition theory is presented in a book by Landau and Lifschitz. Applications of elementary catastrophe theory to phase transitions are dealt with in a book by Poston and Stewart and in a paper by Komorowski.

An excellent introduction to the catastrophe optics contains a paper by Berry and Upstill. Information on the saddle node method of computing oscillation integrals can be found in Wyrzykowski's book and in Nayfeh's book.

The heartbeat model and the model of nerve impulse transmission by the nerve axon are described in Zeeman's paper. A description of ecological and population models can be found in a paper by Auslander and Huffaker, as well as in a paper by May. A more comprehensive treatment of sequential models is found in a paper by Otto. The cited paper of Ulam and von Neumann has been ahead of its time — the authors have examined deterministic chaos generating sequential processes using the first computers available.

References

W. I. Arnol'd, "Osobiennosti, bifurkacyi i katastrofy", *Usp. Fiz. Nauk,* **141**, 569 (1983).

W. I. Arnol'd, "Singularities of ray systems", *Proc. Internat. Congr. Math.,* Vol. 1, Warszawa, 1983, p. 27.

D. G. Auslander and C. Huffaker, "Dynamics of interacting populations", *J. Frank. Inst.,* **5**, 297 (1974).

M. V. Berry and C. Upstill, "Catastrophe optics", E. Wolf (Ed.), *Progr. in optics,* **18**, 257 (1980).

M. J. Feigenbaum, "Quantitative universality for a class of non-linear transformations", *J. Stat. Phys.,* **19**, 25 (1978).

A. T. Fomyenko, *Topologicheskiye Variacyonnye Zadachi,* Izd. Mosk. Univ., Moskva, 1984.

Dao Zhong Thi and A. T. Fomyenko, *Minimalnye Povyerchnosti i Problema Plato,* Izd. Nauka, 1987.

J. Komorowski, "Gas–liquid phase transitions and singularities", *Repts Math. Phys.,* **19**, 257 (1984).

J. A. Kravcov and J. I. Orlov, "Katastrophy, volnovye pola, *Usp. Fiz. Nauk,* **141**/4, 591 (1983).

L. D. Landau and R. M. Lifschitz, *Statistical Physics,* Pergamon, London, 1958.

W. P. Maslov, "Non-standard characteristics in asymptotical problems", *Proc. Intern. Congress of Math.,* Vol. 1, Warszawa, 1983, p. 139.

R. M. May, "Simple mathematical models with very complicated dynamics", *Nature,* **261** (June 10), 459 (1976).

A. H. Nayfeh, *Introduction to Perturbation Techniques,* J. Wiley and Sons, New York, 1981.

E. Ott, "Strange attractors and chaotic motions of dynamical systems", *Rev. Mod. Phys.,* **53**, 655 (1981).

T. Poston, "The Plateau problem. An invitation to the whole of mathematics". Summer College on global analysis and its applications. 4 July–25 August 1972. International Centre for theoretical physics. Trieste, Italy 1972.

T. Poston and I. N. Stewart, *Catastrophe Theory and its Applications,* Pitman, London, 1978.

M. Smoluchowski, *Ann. d. Phys.,* **25**, 205 (1908).

R. Thom, "Symmetries gained and lost", in *Mathematical Physics and Physical Mathematics,* K. Maurin, R. Rączka (Eds.), PWN, Warszawa, 1976.

S. M. Ulam and J. von Neumann, "A combination of stochastic and deterministic dynamics", *Bull. Am. Math. Soc.,* **53**/11, 112 (1947).

E. C. Zeeman, "Differential equations for the heartbeat and the nerve impulse", in *Dynamical Systems,* M. M. Peixoto (Ed.), Academic Press, New York–London, 1973.

Chapter 4

Chemical Kinetics

4.1 INTRODUCTION

Chemical systems constitute exceptionally interesting objects of investigation. They are generally represented by non-linear equations of chemical kinetics which implies that in chemical reactions may occur a number of interesting phenomena associated with the change of a stationary state or the dynamics of a process (phenomena of a catastrophe type) resulting from non-linearity of the process. Apparently, a reacting chemical system constitutes one more realization of the process represented by non-linear differential equations. However, there is an appreciable difference between physical and chemical accomplishments of non-linear processes: in the case of a chemical reaction state variables (concentrations of reagent molecules) are directly related to molecules — the only components of the system implementing a non-linear process. Mixing a few readily available substances enables the generation of an elaborate non-linear dynamical system. At the same time, chemical systems containing an extremely large number of state variables, corresponding to various intermediates formed in a reaction, can be obtained; particular state variables may even be considered to be associated with single molecules or simply with the photons taking part in a reaction and entering into very complex interactions. All these state variables actively participate in a non-linear dynamical process.

Different is the case of physical implementation of non-linear processes. For example, an electronic system with current flow described by a non--linear differential equation has, besides flowing current to which an appropriate state variable (for example, the current intensity) is related, necessary electronic devices which are not described by the current flow equations.

The simplicity of chemical systems and the very complicated dynamics of chemical processes cause the mathematical models of chemical reactions to be an important area of applications of non-linear methods of mathematics, including catastrophe theory.

Keeping in mind the applications of catastrophe theory to a description of changes in the stationary state and the dynamics of a chemical reaction we have to define what will be meant by state variables and control parameters of a chemical system. Prior to that, however, information on the procedure of performing a chemical experiment has to be provided. Chemical reactions are carried out in reactors of various types. For our purposes, a classification of reactors into two classes: closed reactors and flow reactors will suffice. In closed reactors a reaction proceeds in a tank of constant volume without replenishing the reagents used in the reaction and without removing the reaction products from the reactor. Carrying out reactions in flow reactors through which a stream of reacting substances flows is advantageous from the standpoint of kinetic studies. For example, a flow system permits to easily investigate the reacting system in the state far from being stationary (far from equilibrium), thus allowing to detect the intermediates present in those states of the reacting system which are attained under flow conditions. Simultaneous feeding of several solutions, mixing the reactants and removing a partially reacted mixture (in general, via one outlet) are possible. In both types of reactors a number of properties of a reaction mixture, such as pH, light absorption, redox potential, changes in temperature and conductance of a solution can be measured. These quantities are state variables; the reagent concentrations may be related to some of them, e.g. the light absorption can be correlated with a concentration of some of the reagents (absorbing light of a specific wavelength) – then the reagent concentration may be considered to be a state variable. Control parameters are usually the initial composition and concentration of solutions introduced to a closed reactor of flowing into a flow reactor, flow rate of reagents through a reactor, pH, temperature, etc. Hence, a selection of state variables and control parameters depends, to some extent, on the experimenter.

In chemical systems one may thus distinguish state variables and control parameters. Chemical reactions have been known from experience to be structurally stable (resistant to small changes in control parameters) under some conditions; it is also known that the change in a reaction character (change in the stationary state or dynamics) may occur upon a continuous variation in control parameters; in other words, the sensitive state may be created in a chemical system. The above remarks substantiate an attempt to apply catastrophe theory to a description of chemical reactions.

The chapter is arranged as follows. First, the fundamental concepts of chemical kinetics will be discussed, followed by kinetic equations describing chemical reactions. The model of chemical reactions being initially considered allows at most bimolecular interactions (collisions) of the reactants and does not explicitly introduce autocatalytic reactions. Subsequently, the method of reducing and simplifying kinetic equations will be presented. The presented discussion will be used to show how autocatalytic or termolecular reactions can be modelled. A qualitative theory of differential equations of chemical kinetics will then be discussed next. Finally we shall demonstrate how the methods of catastrophe theory may be applied to chemical kinetics equations.

4.2 FUNDAMENTAL CONCEPTS OF CHEMICAL KINETICS

From a theoretical viewpoint it is convenient to classify reactions into homogeneous, occurring entirely in one phase (gaseous or liquid), and nonhomogeneous (heterogeneous), proceeding in more than one phase. Heterogeneous reactions appear to always occur at the phase boundary. Further considerations will be confined to homogeneous reactions.

The rate of a homogeneous reaction occurring in a system of defined composition, at constant temperature, is a function of the concentrations of the species involved. In the case of an important class of reactions, further called elementary, the rate of a reaction is simply related to stoichiometric coefficients present in the notation of the reaction. For example, the initial rate of the reaction, r, proceeding in the gaseous phase

$$H_2 + I_2 \rightarrow 2HI \tag{4.1}$$

in the case of high dilution of the reactants with an inert gas is given by the equation

$$r = 1/2 d/dt [HI] = -d/dt [H_2] = -d/dt [I_2] \tag{4.2a}$$

$$r = k [H_2][I_2] \tag{4.2b}$$

the rate of the reaction r being expressed in mole $l^{-1} s^{-1}$ and corresponding to the time rate of change of a number of moles per unit volume of the reacting system (expressed in liters or cubic decimeters); the constant k is so-called rate constant of the reaction, its dimension $(\text{mole}^{-1} l \, s^{-1})$ resulting from the dimension of r and from equation (4.2); brackets denote concentrations expressed in moles per unit volume.

An elementary reaction, for example (4.1), occurs as a result of direct interaction (collision) of molecules of the reactants without the appearance of any other intermediate states. The rate of an elementary reaction is thus determined by the probability of collision of the reactant molecules, in the above example of a H_2 molecule with an I_2 molecule, which gives rise to dependence (4.2b).

It follows from the examination of experimental data that for any reaction of a general form

$$aA + bB + ... \rightarrow lL + mM + ... \tag{4.3}$$

where small letters stand for the respective stoichiometric coefficients, the dependence of the rate of a reaction on the concentrations may be generally expressed by a simple product equation

$$r = k[A]^\alpha [B]^\beta ... [L]^\lambda [M]^\mu ... \tag{4.4}$$

where k is as in (4.2b), the rate constant of a reaction, dependent only on the temperature. Then, the constant α is the reaction order with respect to A, β is the reaction order with respect to B, etc., and the overall reaction order is the sum of exponents in equation (4.4)

$$\text{the order of reaction (4.3)} = \alpha + \beta + ... + \lambda + \mu + ... \tag{4.5}$$

In cases when the rate of a reaction cannot be expressed by equation (4.4) and depends on the concentration of the reagents in a more elaborate way, the notion of the reaction order obviously loses its meaning.

An important concept in chemical kinetics is molecularity of a reaction or the number of particles (molecules, atoms, ions, radicals) participating in it. Most common are bimolecular reactions, unimolecular reactions being also encountered. In very rare cases termolecular reactions may be observed as well. Reactions of higher molecularity are unknown, which is due to a very low probability of a simultaneous interaction of a larger number of molecules. Consequently, our further considerations will be confined to the examination of uni- and bimolecular reactions. On the other hand, the reactions of a termolecular character, whose kinetic equations have a number of interesting properties, are sometimes considered. As will appear, a termolecular reaction may be approximately modelled by means of a few bimolecular reactions. For an elementary reaction its molecularity is by definition equal to the order whereas for a complex reaction the molecularity generally has no relation whatsoever to the reaction order or the stoichiometry.

The exponents in equation (4.4) may occasionally be equal to the stoichiometric coefficients in the reaction equation ($\alpha = a$, $\beta = b$, etc.). Apparently, that is the case for elementary reactions, for example cf. (4.1), (4.2b). In the case of a general reaction, for which (4.4) holds, this is no longer valid. This is so because in general a reaction consists of a sequence of elementary reactions which constitute the so-called mechanism of the reaction. In such a case, an equation for the rate of a reaction of the form (4.4) cannot be deduced without knowledge of the mechanism of the reaction (the overall chemical equation does not reflect elementary acts of the interaction of molecules — the collisions).

An example of a complex (nonelementary) reaction may be the reaction of Br^- with BrO_3^- yielding Br_2;

$$BrO_3^- + 5\,Br^- + 6\,H^+ \rightarrow 3\,Br_2 + 3\,H_2O \qquad (4.6)$$

The rate of this reaction may be written as the rate of decrease of Br^- or BrO_3^- ions, or the rate of Br_2 formation, (t stands for time):

$$r = -d/dt\,[BrO_3^-] = -1/5\,d/dt\,[Br^-] = 1/3\,d/dt\,[Br_2] \qquad (4.7)$$

The initial rate of Br_2 formation has been experimentally determined to be

$$r = k\,[BrO_3^-][Br^-][H^+]^2, \; k = 2.1\,(\text{mole}^{-3}\,l^3\,s^{-1}) \qquad (4.8)$$

the rate constant k of the reaction having been measured at 25°C. The dependence (4.8) may be explained by the following sequence of elementary reactions (the mechanism)

$$BrO_3^- + Br^- + 2\,H^+ \rightarrow HBrO_2 + HOBr \qquad (4.9a)$$

$$HBrO_2 + Br^- + H^+ \rightarrow 2\,HOBr \qquad (4.9b)$$

$$HOBr + Br^- + H^+ \rightarrow Br_2 + H_2O \qquad (4.9c)$$

in which each of the elementary reactions involves the transfer of one oxygen atom from one molecule to another in the act of an elementary collision. The experimentally measured rates of the reactions (4.9) confirm their elementary nature

$$r_a = k_a\,[BrO_3^-][Br^-][H^+]^2, \quad k_a = 2.1\,(\text{mole}^{-3}\,l^3\,s^{-1}) \qquad 4.10a)$$

$$r_b = k_b\,[HBrO_2][Br^-][H^+], \quad k_b = 2\cdot10^9\,(\text{mole}^{-3}\,l^3\,s^{-1}) \qquad (4.10b)$$

$$r_c = k_c\,[HOBr][Br^-][H^+], \quad k_c = 8\cdot10^9\,(\text{mole}^{-3}\,l^3\,s^{-1}) \qquad (4.10c)$$

as the exponents in (4.10) are equal to the respective stoichiometric coefficients in equations (4.9). Reaction (4.6) may also be written as $(6) = (9a) + (9b) + 3(9c)$.

When a reaction proceeds through several successive elementary steps, and one of these reactions is very much slower than any of the others, then the rate will depend on the rate of this single slowest step. The slow step is the rate-determining step. In the above example, reaction (9a) constitutes such a rate-determining step. In reactions (9a) and (9b), a steady state for $[HBrO_2]$ is rapidly reached (so-called pseudosteady state). $HBrO_2$ formed in reaction (9a) is almost immediately consumed in reaction (9b). Hence, the formation of HOBr in (9b) occurs at the same rate as in (9a). From the condition of equality of the rates of the two reactions in a pseudosteady state, $r_a = r_b$, we obtain

$$k_a[BrO_3^-][Br^-][H^+]^2 = k_b[HBrO_2][Br^-][H^+] \tag{4.11}$$

Likewise, a pseudosteady state is rapidly reached for [HOBr] in reactions (9a), (9c), i.e. from the condition $r_a = r_c$ we have

$$k_a[BrO_3^-][Br^-][H^+]^2 = k_c[HOBr][Br^-][H^{+2}] \tag{4.12}$$

It follows from the above considerations that in the sequence of reactions (4.9) a pseudosteady state is really reached and its properties derive from equations (4.11), (4.12):

$$[HBrO_2] = k_a/k_b[BrO_3^-][H^+] \tag{4.13a}$$

$$[HOBr] = k_a/k_b[BrO_3^-][H^+] \tag{4.13b}$$

Thus, the sequence of reactions (4.9) may be effectively replaced by reaction (4.9a) with conditions (4.13). We may then forget that reactions (4.9b), (4.9c) also occur − in a time scale determined by reaction (4.9a) they proceed all but instantaneously and their effects are expressed to a good approximation by equations (4.13). The rate of nonelementary reaction (4.6) is thus consistent with the experimental equation (4.8). By a time scale characteristic of a given reaction we will mean a time interval after which concentrations of reagents undergo a distinct change.

It follows from the principles of thermodynamics that all chemical reactions are reversible; this fact has so far been omitted. In some cases the effects associated with reversibility of chemical reactions have to be taken into account. For example, the reaction reverse to (4.9c):

$$Br_2 + H_2O \rightarrow HOBr + H^+ \tag{4.14}$$

proceeds at the rate

$$r_{-c} = k_{-c}[Br_2], \quad k_{-c} = 10^2 \, (s^{-1}) \tag{4.15}$$

where the fact that in dilute aqueous solutions $[H_2O] = 55.5 \, (\text{mole } l^{-1})$ was taken into account; this constant was incorporated into k_{-c}. Accounting for the occurrence of the reverse reaction, the net rate of changes in $[Br_2]$, resulting from reactions (4.9c), (4.14), is equal to:

$$d/dt[Br_2] = r_c - r_{-c} = k_c[HOBr][Br^-][H^+] - k_{-c}[Br_2] \tag{4.16}$$

When (4.9c), (4.14) reach an equilibrium state, the condition

$$d/dt[Br_2] = 0 \tag{4.17}$$

is satisfied. Then it follows from equation (4.16) that

$$\frac{[Br_2]}{[HOBr][Br^-][H^+]} = \frac{k_c}{k_{-c}} \equiv K_c, \quad K_c = 8 \cdot 10^7 \, (\text{mole}^{-2} \, l^2) \tag{4.18}$$

The constant $K_c = k_c/k_{-c}$ is called the equilibrium constant of reactions (4.9c), (4.14).

4.3 KINETIC EQUATIONS FOR REACTIONS WITHOUT DIFFUSION

4.3.1 Examples of kinetic equations without diffusion

On the basis of known mechanism of a reaction, kinetic equations describing changes in concentration of reagents taking place during the reaction can be written. When the effects related to diffusion are not taken into consideration we speak, none too precisely, about reaction equations without diffusion (a so-called point system of kinetic equations) whereas in the case of accounting in kinetic equations for diffusion we deal with reaction equations with diffusion (a so-called non-point system of equations). In the present section we shall discuss properties of equations of reactions without diffusion. The equations of reactions with diffusion and the problem of a correct limit transition from equations with diffusion to diffusionless equations will be examined in the following section.

Consider an example describing a sequence of isomerization reactions:

$$A \underset{k_{-1}}{\overset{k_1}{\rightleftharpoons}} B \underset{k_{-2}}{\overset{k_2}{\rightleftharpoons}} C \tag{4.19}$$

The concentration of B increases at the rate k_1 [A] and decreases at the rate k_{-1} [B] as a result of the first reaction, and decreases at the rate k_2 [B] and increases at the rate k_{-2} [C] as a result of the second reaction. The changes in [A] and [C] may be expressed in a similar way.

Introducing designations $a =$ [A], $b =$ [B], $c =$ [C], we thus arrive at the following system of equations describing the changes in [A], [B], [C] in the course of the reaction:

$$da/dt = k_1 a + k_{-1} b \tag{4.20a}$$

$$db/dt = k_1 a - (k_{-1} + k_2)b + k_{-2} c \tag{4.20b}$$

$$dc/dt = k_2 b - k_{-2} c \tag{4.20c}$$

Summation of equations (4.20) yields

$$d/dt (a + b + c) = 0 \tag{4.20d}$$

that is, the law of conservation of mass

$$a + b + c = \text{const} \tag{4.20d'}$$

The system of equations (4.20) falls into a class of systems of ordinary differential equations: it is an autonomous dynamical system, see equation (1.6) in Section 1.2. Furthermore, system (4.20) is linear — all unknowns are in the first power. As will be shown later, the systems of equations of this type may be readily solved. Linearity of the system of kinetic equations is a consequence of unimolecularity of elementary reactions in the mechanism (4.19).

By and large, reacting systems cannot be described by linear differential equations. In typical cases, quadratic terms resulting from bimolecular interactions of reagents occur in equations (molcularity of a reaction equal to two). For example, the mechanism of the Edelstein enzymatic reaction is of the form

$$A + X \underset{k_{-1}}{\overset{k_1}{\rightleftharpoons}} 2X \tag{4.21a}$$

$$X + \underset{k_{-2}}{\overset{k_2}{\rightleftharpoons}} D \tag{4.21b}$$

$$D \underset{k_{-3}}{\overset{k_3}{\rightleftharpoons}} Y + F \tag{4.21c}$$

where (4.21a) describes an autocatalytic production of X from A while (4.21b), (4.21c) represent an enzymatic degradation of X to F catalyzed by the enzyme Y and proceeding with regeneration of the enzyme Y. The intermediate D is a complex, D = XY. The total enzyme concentration, e, is thus given by

$$e = [Y] + [D] \equiv y + d \tag{4.21d}$$

Let us assume that during reaction (4.21) the concentrations $[A] \equiv a$, $[F] \equiv f$ and $[Y] + [D]$ are maintained at a constant level:

$$a = \text{const}, \quad f = \text{const}, \quad e = y + d \tag{4.21e}$$

and only $x \equiv [X]$, $y \equiv [Y]$, and $d \equiv [D]$ undergo changes. Hence, changes in independent concentrations x, y (d is related to y via equation (4.21e) are described by the following system of equations:

$$dx/dt = k_1 ax - k_{-1}x^2 - k_2xy + k_{-2}y \tag{4.22a}$$

$$dy/dt = -k_2xy - (k_2 + k_3)d + k_{-3}fy \tag{4.22b}$$

with conditions (4.21e).

The system of equations (4.22) is a non-linear system of ordinary differential equations; non-linearities derive from bimolecular steps of reactions (4.21a)–(4.21c).

Consider one more mechanism of a reaction investigated by the Prigogine school (the model is called Brusselator)

$$A \overset{k_1}{\longrightarrow} X \tag{4.23a}$$

$$B + X \overset{k_2}{\longrightarrow} Y + D \tag{4.23b}$$

$$2X + Y \overset{k_3}{\longrightarrow} 3X \tag{4.23c}$$

$$X \overset{k_4}{\longrightarrow} E \tag{4.23d}$$

We assume that all the reactions are practically irreversible, the concentrations $a = [A]$, $b = [B]$ are maintained at a constant level, the concentrations of the final products $d = [D]$, $e = [E]$ do not affect the reaction kinetics, the concentrations of the intermediates $x = [\mathrm{X}]$, $y = [\mathrm{Y}]$ change in the course of the reaction.

For the above assumptions, equations for changes in concentration of the intermediates with time have the following form:

$$\mathrm{d}x/\mathrm{d}t = k_1 a - k_2 bx + k_3 x^2 y - k_4 x \tag{4.24a}$$

$$\mathrm{d}y/\mathrm{d}t = k_2 bx - k_3 x^2 y \tag{4.24b}$$

On introducing the dimensionless quantities

$$t' = k_4 t \tag{4.25a}$$

$$x' = (k_3/k_4)^{1/2} x, \qquad y' = (k_3/k_4)^{1/2} y \tag{4.25b}$$

$$a' = (k_1/k_4)(k_3/k_4)^{1/2} a, \qquad b' = (k_2/k_4) b \tag{4.25c}$$

system (4.24) may be written as

$$\mathrm{d}x'/\mathrm{d}t' = a' - b'x' + x'^2 y' - x' \tag{4.26a}$$

$$\mathrm{d}y'/\mathrm{d}t' = b'x' - x'^2 y' \tag{4.26b}$$

Once again, a non-linear system of ordinary differential equations is obtained, the cubic term $x'^2 y'$ appearing as a consequence of the termolecular step of the reaction, (4.23c).

4.3.2 Standard kinetic systems

There exist theoretical grounds for the derivation of all kinetic equations from the so-called standard kinetic systems. Standard systems have the following form:

$$\dot{x}_i = \sum_{k=1}^{n} b^k x_k + \sum_{k,l=1}^{n} c^{kl} x_k x_l, \quad i = 1, 2, ..., n$$

$$b_i^i \leqslant 0, \quad c_i^{il} \leqslant 0, \quad c_i^{ki} \leqslant 0 \quad x_i \geqslant 0 \tag{4.27}$$

where b^k, c^{kl} are numerical factors, $\dot{x}$ denotes $\mathrm{d}x/\mathrm{d}t$.

Thus, standard systems have the following characteristics:

(a) at most bimolecular interactions of reagents are taken into account since, as already mentioned in Section 4.2, the interactions of larger molecularity have low probability;

(b) the coefficients b_i^i, c_i^{il}, c_i^{ki} are non-positive (the remaining coefficients have arbitrary signs) which implies that the reaction mechanism does not contain autocatalytic elementary reactions (each elementary reaction results in a decrease in concentration of reactants);

(c) all quantities x_i represent concentrations and are thus non-negative.

Hence, restriction of considerations to a class of standard systems (4.27) has a theoretical substantiation.

The system (4.27) may fulfil the law of conservation of mass

$$\sum_{i=1}^{n} d_i x_i = \text{const} \tag{4.28}$$

where d_i stands for the molecular weight of a substance x_i; such a system is called a closed system.

According to the above terminology the system (4.20) is a closed standard system, the system (4.22) is an open autocatalytic $(b^1 = k_1 > 0)$ system, the system (4.26) is not a standard system also, because it is open and contains the third-order term $x^2 y$; moreover, for suitable values of the control parameters a', b' it can be an autocatalytic system.

Thus, the problem of exactness of kinetic systems which are not standard, for example (4.22) or (4.26), seems to appear. As will be shown, the systems containing autocatalytic terms or termolecular interactions may be modelled, with a desired accuracy, be means of the standard system (4.27).

There still remains the problem of description of open systems in which mass is exchanged with the surroundings; then, the condition (4.28) is not satisfied. Two approaches are possible in this case. A suitably large reserve of molecules (for example A, B in equations (4.19), (4.20)) may be introduced, thereby making the system formally closed (the concentrations a, b may then be considered constant for an appropriately long time). The terms describing exchange of mass with the surroundings (flow of mass through a reactor) may also be included into the system of equations (4.27):

$$\dot{x}_i = \sum_{k=1}^{n} b^k x_k + \sum_{k,l=1}^{n} c^{kl} x_k x_l + 1/T \left(x_i^{\,0} - x_i \right) \tag{4.29}$$

where T is the retention time of a flow reactor (the average residence time of a substance in the reactor), $x_i^{\,0}$ and x_i are the concentrations of a given substance in a stream flowing into and leaving the reactor, respectively.

The system (4.29) obviously fulfils the law of conservation of mass, since the entire mass in a stream entering the reactor and in a stream leaving it is preserved.

4.4 KINETIC EQUATIONS FOR REACTIONS WITH DIFFUSION

Note that ignoring the effects associated with diffusion of substances reacting in solution requires substantiation, since the contact of reagent molecules is achieved due to stirring and diffusion.

Consider a diffusionless equation describing an evolution of concentration of two substances X, Y, $[X] \equiv x$, $[Y] \equiv y$ with time:

$$dx/dt = P(x, y) \tag{4.30a}$$

$$dy/dt = Q(x, y) \tag{4.30b}$$

where P, Q are some generally non-linear functions x, y.

Let us assume that a reaction is carried out in a reactor shaped as a thin, long pipe.

Let us now allow the possibility of diffusion of the molecules X, Y along the reactor axis (the diffusion in lateral directions may be neglected). The variables x, y, functions of time t, also become dependent on the distance r from one of the reactor walls, measured along its axis. The terms representing transport of the substances X, Y due to diffusion should be introduced into equations (4.30). Recall that the non-dimensional diffusion equation is of the form:

$$\partial f(t, r)/\partial t = \partial/\partial r [D_r \partial f(t, r)/\partial r) \tag{4.31}$$

where D_r is the diffusion coefficient of the substance f. The diffusion coefficient may be frequently assumed to be constant:

$$\partial f(t, r)/\partial t = D_r \partial^2 f(t, r)/\partial r^2 \tag{4.32}$$

Thus, to account for diffusion in equations (4.29), they should be modified as follows:

$$\partial x(t, r)/\partial t = P[x(t, r), y(t, r)] + D_x \partial^2 x(t, r)/\partial r^2 \tag{4.33a}$$

$$\partial y(t, r)/\partial t = Q[x(t, r), y(t, r)] + D_y \partial^2 y(t, r)/\partial r^2 \tag{4.33b}$$

The system (4.30) may be formally obtained from (4.33) by substituting $D_x = D_y = 0$. However, this would lead, under no stirring, to an inability of contact of the reagents and to termination of a chemical reaction. One may pass on, however, from the system (4.33) to (4.30) by way of a more correct procedure, consisting in such a rescaling of variables that the dimensions of the system along the r-axis may be ignored and then the terms representing diffusion may also be neglected (as we did with the diffusion in directions perpendicular to the reactor axis). Since due to such a limiting procedure the system contracts to a point, the systems of type (4.30) are sometimes called point systems.

The rescaling of the variable r mentioned above is of the form

$$r \to r' = r/D, \quad D = (D_x D_y)^{1/2} \tag{4.34}$$

In new variables t, r equations of the system (4.33) become

$$\partial x(t, r')/\partial t = P(x, y) + (D_x/D^2)\partial^2 x(t, r')/\partial r'^2 \tag{4.35a}$$

$$\partial y(t, r')/\partial t = Q(x, y) + (D_y/D^2)\partial^2 y(t, r')/\partial r'^2 \tag{4.35b}$$

At the limit D_x, $D_y \to \infty$ the system (4.35) is transformed into the diffusionless system (4.30). If the reactor length in old variables was l_0, then in new variables it amounts to $l_1 = l_0/D$. The limiting transition $(D_x, D_y \to \infty)$ implies an infinitely large diffusion coefficient of the substances X, Y and zero volume of the reactor. It follows from the above analysis that the diffusional terms may be disregarded when the diffusion processes are so effective (large D_x and D_y) that, due to occurring chemical reactions, the concentrations of the reagents may be considered to change synchronously along the entire length of the reactor. It should also be taken into consideration that stirring aids the diffusion processes.

4.5 MODELLING OF THE REACTION MECHANISMS BY STANDARD KINETIC SYSTEMS

4.5.1 Introduction

In numerous cases the standard system of kinetic equations (4.27) may be effectively replaced with a considerably simpler system of equations. This results in a number of advantages. Firstly, the reduced system of equations is easier to examined. Secondly, as we will show later, such a reduced system may contain autocatalytic terms and non-linear terms of order higher than two. Thirdly, the method of reduction of a standard system enables a direct transposition of the methods of elementary catastrophe theory to the ground of chemical kinetics equations.

4.5.2 Fundamentals of the Tikhonov's method of reduction of kinetic equations

The principle of the Tikhonov (steady concentrations) method derives from an observation that in a set of reactions leading to equations (4.27), rapid and slow reactions can be distinguished. This means that the system (4.27) may be written in the following form:

$$\varepsilon^2 dx_i/dt = F_i(x_1, ..., x_n), \quad i = 1, ..., l < n \tag{4.36a}$$

$$\varepsilon dx_j/dt = F_j(x_1, ..., x_n), \quad j = l + 1, ..., l + m < n \tag{4.36b}$$

$$dx_k/dt = F_k(x_1, ..., x_n), \quad k = l + m + 1, ..., n \tag{4.36c}$$

The coefficients ε^2, ε, l determine the rate of changes in concentrations x_i, x_j, x_k and define time scales characteristic of the processes (4.36a), (4.36b), (4.36c). We may introduce new variables T'', T', T,

$$\varepsilon^2 dx_i/dt \equiv dx_i/dT'', \quad T'' = \varepsilon^{-2}t \tag{4.37a}$$

$$\varepsilon dx_j/dt \equiv dx_j/dT', \quad T' = \varepsilon^{-1}t \tag{4.37b}$$

$$dx_k/dt \equiv dk_k/dT, \quad T = t \tag{4.37c}$$

corresponding to different time scales. Equations (4.36) are divided into three groups of equations evolving in different time scales:

$$dx_i/dT'' = F_i(x_1, ..., x_n), \quad i = 1, ..., l < n \tag{4.38a}$$

$$dx_j/dT' = F_j(x_1, ..., x_n), \quad j = l + 1, ..., l + m < n \tag{4.38b}$$

$$dx_k/dT = F_k(x_1, ..., x_n), \quad k = l + m + 1, ..., n \tag{4.38c}$$

If the parameter ε is small, $\varepsilon < 1$, then equations (4.36a) represent rapid processes, equations (4.36c) describe slow processes and equations (4.36b) correspond to processes having intermediate rates of changes with time.

In the case when we are interested mainly in the evolution of variables x_j with time, that is the overall process is examined in the time scale T', the system of equations (4.36) may substantially simplified. It is intuitively clear that if the functions F_i, F_j, F_k do not have pathological properties, the slow variables $x_k = (x_{l+m+1}, ..., x_n)$ may be regarded as constants in the time scale T' whereas the variables $x_i = (x_1, ..., x_l)$, evolving rapidly, may be replaced with the equilibrium values attained by them (rapidly in the time scale T').

The conditions for which a reduction of the system (4.36) described above is allowed, are given by the Tikhonov theorem.

4.5.3 Tikhonov theorem

Let a system of n kinetic equations be of the form

$$\varepsilon \, dx_i/dt = F_i(x_1, ..., x_l, ..., x_n), \quad i = 1, ..., l < n \tag{4.39a}$$

$$dx_j/dt = F_j(x_1, ..., x_l, ..., x_n), \quad j = l+1, ..., n \tag{4.39b}$$

where ε is a small parameter, $\varepsilon \ll 1$. The system of equations (4.39a) will be called an adjoined (rapid) system and the system of equations (4.39b) will be referred to as a degenerate (slow) system.

A solution to the overall system (4.39) may be proved to approach, for $\varepsilon \to 0$, the solution to the following system of equations:

$$F_i(x_1, ..., x_l, ..., x_n) = 0, \qquad i = 1, ..., l \tag{4.40a}$$

$$dx_j/dt = F_j(x_1, ..., x_l, ..., x_n), \quad j = l+1, ..., n \tag{4.40b}$$

if the following requirements are satisfied:

(a) a solution to the system of algebraic equations (4.40a):

$$\bar{x}_1 = f_1(x_{l+1}, ..., x_n), ..., \qquad \bar{x}_l = f_l(x_{l+1}, ..., x_n) \tag{4.41}$$

is an isolated root (that is, there are no other solutions in an arbitrarily small neighbourhood of the point $(\bar{x}_1, ..., \bar{x}_l)$. The system of equations (4.40b) is solved on substituting into it the solutions to equations (4.40a):

$$dx_j/dt = F_j(\bar{x}_1, ..., \bar{x}_l, x_{l+1}, ..., x_n), \quad j = l+1, ..., n \tag{4.40b'}$$

(b) the solution $\bar{x}_1(x_{l+1}, ..., x_n), ..., \bar{x}_l(x_{l+1}, ..., x_n)$ is a stable isolated stationary point of the rapid system (4.39a) for all the values $x_{l+1}, ..., x_n$ (a stationary point meets the requirement $dx_i/dt = 0$);

(c) the initial values $x_1^0, ..., x_l^0$ are in the attracting region of a stationary point of the rapid system (4.39a);

(d) the solutions to the systems of equations (4.39), (4.40) are unique and the functions occurring on the right-hand side of these systems are continuous.

Conditions (a), (c), (d) are generally fulfilled in chemical systems whereas condition (b) may be unsatisfied in a case of processes in which oscillations in concentration occur. In accordance with our general methodology, the situation wherein with a continuous variation in control parameters condition (b) ceases to be met may be regarded as a catastrophe. The examination of catastrophes of this type (in general, these are not elementary catastrophes) will be described in Chapter 6.

The Tikhonov theorem has an important generalization, called the centre manifold theorem, which will be discussed in Sections 5.4.5–5.4.7. In classification of catastrophes occurring in dynamical systems and represented by systems of autonomous equations, the centre manifold theorem plays the role of the splitting lemma (see Section 2.3.4).

4.5.4 Application of the Tikhonov theorem to modelling of dynamical systems and chemical reactions

Using the Tikhonov theorem, the desired behaviour of dynamical system may be modelled or, alternatively, one may obtain from the standard system of chemical kinetics equations (4.27) effective systems of equations which cannot be represented in this form (for example, of an autocatalytic type or termolecular).

In Section 3.5 we have described the Zeeman models for the heartbeat and the nerve impulse transmission in terms of systems of ordinary differential equations (dynamical systems). The idea of modelling these systems boiled down to designing a dynamical system having two different time scales. Slow dynamics of a suitable model and the following catastrophe proceeding according to fast dynamics satisfied some general conditions imposed by experiment. From a technical standpoint, the process of modelling was reduced to an application of the Tikhonov theorem to an increasingly more elaborate dynamical system in order to find the shape of a slow surface, determining the properties of the model.

Standard kinetic systems whose slow dynamics has a desirable nature may be designed in a similar way. In Section 4.5.5 we shall answer the question whether the standard system (4.27) may have arbitrarily slow dynamics.

We will now employ the Tikhonov theorem to designing standard systems whose slow dynamics is of an autocatalytic or termolecular character. We shall begin with modelling an autocatalytic reaction (such as, for example, Edelstein reaction (4.21)).

Consider the reaction mechanism consisting of two elementary reactions in which occur two reactants A and B, intermediates X_1 and X_2 and a final product Z:

$$A + X_1 \xrightarrow{k_1} 2X_2 \tag{4.42a}$$

$$B + X_2 \xrightarrow{k_2} X_1 + Z \tag{4.42b}$$

assuming that the effect of reverse reactions and the effects associated with diffusion may be ignored $(k_{-1} = 0,\ k_{-2} = 0$, see also the discussion in Section 4.4). With these assumptions the system of kinetic equations, corresponding to reactions (4.42), has the form

$$\dot{a} = -k_1 a x_1, \quad \dot{b} = -k_2 b x_2, \quad \dot{z} = k_2 b x_2$$

$$\dot{x}_1 = -k_1 a x_1 + k_2 b x_2, \quad \dot{x}_2 = 2k_1 a x_1 - k_2 b x_2 \tag{4.43}$$

small letters denoting concentrations of the respective substances, $a \equiv [A]$, $b \equiv [B]$, etc. By adding the two equations (4.43) we obtain

$$d/dt(a + b + x_1 + x_2 + z) = 0 \tag{4.44}$$

which implies that the system (4.43) is a closed standard system.

Let us assume that the reserves of the substances A, B are so large that the concentrations a, b are practically constant throughout the reaction. For these assumptions, the system of kinetic equations (4.43) is reduced to equations representing the time evolution of x_1, x_2:

$$\dot{x}_1 = -k_1 a x_1 + k_2 b x_2 \tag{4.45a}$$

$$\dot{x}_2 = 2k_1 a x_1 - k_2 b x_2 \tag{4.45b}$$

The system of equations (4.45) is linear and may be readily solved; however, we will apply the Tikhonov theorem to show how to use it. Assume also that the following relationship

$$k_1 a / k_2 b = \varepsilon \ll 1 \tag{4.46}$$

is valid.

Summing equations (4.45) yields

$$d/dt(x_1 + x_2) = k_1 a x_1 = \varepsilon(k_2 b x_1) \tag{4.47}$$

and thus (4.45) is approximately a closed system.

Upon introducing a new time scale

$$T' = k_1 a t \tag{4.48}$$

the system (4.45) is written in the form

$$dx_1/dT' = -x_1 + \varepsilon^{-1} x_2 \tag{4.49a}$$

$$dx_2/dT' = 2x_1 - \varepsilon^{-1} x_2 \tag{4.49b}$$

Introduction of new variables $y_1 = \varepsilon x_1$, $y_2 = x_2$ allows to write down the system in the form

$$dy_1/dT' = -y_1 + y_2 \tag{4.50a}$$

$$\varepsilon dy_2/dT' = 2y_1 - y_2 \tag{4.50b}$$

We may now apply the Tikhonov theorem to the system (4.50): equation (4.50b) is a rapid system whereas (4.50a) is a slow equation and describes slow processes.

Application of the Tikhonov theorem leads to the system of equations

$$dy_1/dT' = -y_1 + y_2 \tag{4.51a}$$

$$0 = 2y_1 - y_2 \tag{4.51b}$$

in which the second equation, $y_2 = 2y_1$, is an equation of a slow surface. Substitution of this equation into (4.50a) yields an effective equation describing slow dynamics of the system (4.50) or (4.45) with the assumption (4.46):

$$dy_1/dT' = +y_1 \tag{4.52}$$

and, after substituting the definition of y_1, $y_1 = \varepsilon x_1$, we obtain

$$dx_1/dT' = +x_1 \tag{4.53}$$

that is, the kinetic system of an autocatalytic nature. The solution to this equation is of the form $x_1(T') = \text{const } \exp(T')$.

Hence, it appears that, on satisfying the condition (4.46), the standard system (4.45) may be replaced by the effective equation (4.53) — slow dynamics of the initial system has an autocatalytic character. Equations

(4.51) are a good approximation of the system (4.45) as long as the changes in concentration of the components A and B may be ignored.

As a second application of the Tikhonov theorem we will present the derivation of an effective kinetic equation containing the termolecular interaction from a standard system of kinetic equations. Such an interaction occurs in the sequence of reactions (4.23) and in the system of kinetic equations the cubic term $x^2 y$ is present. We shall demonstrate that such a term may appear in slow dynamics of a system of elementary reactions at most bimolecular (a standard system). Consider the following sequence of elementary reactions

$$X + Y \underset{k_1}{\overset{k_1}{\rightleftharpoons}} A \tag{4.54a}$$

$$A + Y \overset{k_2}{\longrightarrow} 3X \tag{4.54b}$$

$$\text{(slow processes)} \tag{4.54c}$$

in which the processes (4.54a), (4.54b) are fast, and (4.54c) denotes slow processes with the participation of substances X, Y whereas substance A does not take part in them.

The mechanism (4.54) leads to the standard system of kinetic equations,

$$\dot{x} = -k_1 xy + k_{-1} a + 3k_2 ay + \{...\} \tag{4.55a}$$

$$\dot{y} = -k_1 xy + k_{-1} a - k_2 ay + \{...\} \tag{4.55b}$$

$$\dot{a} = k_1 xy - k_{-1} a - k_2 ay \tag{4.55c}$$

where $x \equiv [X]$, $y \equiv [Y]$, $a \equiv [A]$ and the expressions in braces denote contributions from slow processes.

Applying now the Tikhonov theorem, fast equation (4.55c) may be replaced with

$$0 = k_1 xy - k_{-1} a - k_2 ay \tag{4.56}$$

and, consequently, after computing a from (4.56) the remaining equations can be written in the following form:

$$\dot{x} = -k_1 xy + \left(k_{-1} + 3k_2 y\right) k_1 xy / \left(k_{-1} + k_2 y\right) + \{...\} \tag{4.56a}$$

$$\dot{y} = -k_1 xy + \left(k_{-1} - k_2 y\right) k_1 xy / \left(k_{-1} + k_2 y\right) + \{...\} \tag{4.56b}$$

The fraction $1/(k_{-1} + k_2 y)$ may be expanded in a geometric series in y. If $k_{-1} > k_2$, then the series is convergent. The first term of the series represents in equations (4.56), describing slow dynamics of the system (4.55), terms proportional to $x^2 y$.

Note also that the sequence of reactions (4.54) may be written in a more realistic form. For example, reactions (4.54a), (4.54b) may be replaced with the elementary bimolecular reactions

$$X + F \underset{k_{-1a}}{\overset{k_{1a}}{\rightleftharpoons}} XF \tag{4.57a}$$

$$XF + Y \underset{k_{-1b}}{\overset{k_{1b}}{\rightleftharpoons}} XYF \tag{4.57b}$$

$$XYF + Y \overset{k_2}{\longrightarrow} XY_2F \tag{4.57c}$$

$$XY_2F \overset{k_3}{\longrightarrow} 3X + F \tag{4.57d}$$

where F is an enzyme and the complexation processes proceed rapidly.

4.5.5 Slow dynamical systems and chemical kinetics equations

The desired behaviour of a chemical dynamical system can be modelled by an effective system of kinetic equations in the way similar to that described in Section 3.5 for modelling the heartbeat. The method involves designing a system of differential equations having the desired slow dynamics (the proper slow surface). We should now answer the question whether application of the Tikhonov theorem to the standard kinetic system (4.27) may yield a completely arbitrary slow dynamical system (4.40b'). A partial answer to this question is provided by the Korzukhin theorem: Each dynamical system of the form

$$\left.\begin{aligned} \dot{x}_1 &= F_1(x_1, ..., x_m) \\ &\cdots\cdots\cdots\cdots\cdots\cdots\cdots \\ \dot{x}_m &= F_m(x_1, ..., x_m) \end{aligned}\right\} \quad \begin{aligned} &F_1, ..., F_m \text{ are any polynomials} \\ &\text{in variables } x_1, ..., x_m \end{aligned} \tag{4.58}$$

may be approximated with an arbitrary accuracy by a certain system of elementary reactions at most bimolecular (to which corresponds the standard system of kinetic equations (4.27)). Then, the system (4.58) corresponds to slow dynamics of a certain standard system.

From the Korzukhin theorem follows an important conclusion. Any dynamical systems of the form (4.58) may be regarded as those corresponding to slow dynamics of a standard kinetic system. In other words, the behaviour of dynamical systems can be modelled using chemical reactions. In particular, any of the gradient systems may be modelled in this way. As will be shown in Chapter 5, catastrophes occurring in complex dynamical systems are equivalent to catastrophes appearing in much simpler systems. The latter can be classified — these are so-called standard forms. The standard forms are of the form (4.58) and it follows from the Korzukhin theorem that they can be modelled by the standard equations of chemical kinetics (4.27), corresponding to a realistic mechanism of chemical reactions.

Bibliographical Remarks

Fundamental information on chemical kinetics and kinetics equations can be found in the books of Avery, or Berry, Rice and Ross, as well as books by Tyson and Murray.

Much information on kinetic equations and their reduction using the Tikhonov theorem and catastrophe theory is provided in a book by Romanovskii, Stepanova and Chernavskii.

An additional material concerning the relation of the Tikhonov theorem with catastrophe theory is presented in a paper by Feinn and Ortoleva.

References

H. E. Avery, *Basic Reaction Kinetics and Mechanisms*, MacMillan, 1974.

R. S. Berry, S. A. Rice and J. Ross, *Physical Chemistry*, Part three, *Physical and Chemical Kinetics*, J. Wiley and Sons, New York–Chichester–Brisbane–Toronto, 1980.

W. Ebeling, *Strukturbildung bei irreversiblen Prozessen*, BSB B. G. Teubner Verlagsgesellschaft, Leipzig, 1976.

D. Feinn and P. Ortoleva, "Catastrophe and propagation in chemical reactions", *J. Chem. Phys.*, **67**, 2119 (1977).

R. J. Field and M. Burger (Eds.), *Oscillations and Travelling Waves in Chemical Systems*, J. Wiley and Sons, New York–Chichester–Brisbane–Toronto–Singapore, 1985.

H. Haken, *Synergetics*, Springer-Verlag, Berlin–Heidelberg–New York, 1978.

J. D. Murray, *Lectures on Nonlinear-Differential-Equation Models in Biology*, Clarendon Press, Oxford, 1977.

J. M. Romanovskii, N. W. Stepanova and D. S. Chernavskii, *Mathematical Biophysics* (in Russian), Nauka, Moskva 1984.

J. J. Tyson, *The Belousov–Zhabotinskii Reaction*, Springer-Verlag, Berlin–Heidelberg–New York, 1976, Lecture Notes in *Biomathematics*, **10**, 1 (1976).

Chapter 5

Catastrophes in Dynamical Systems

5.1 INTRODUCTION

The discussion of catastrophes occurring in dynamical systems will be preceded by a description of the classification of linear systems and some of the properties of non-linear systems. The relation between properties of non-linear systems and linear systems obtained by their linearization (the Grobman–Hartman theorem) will be given. Then, we will show how the notions of elementary catastrophe theory are transferred to dynamical gradient systems. In the next section nongradient systems, and particularly analogies to gradient systems, will be briefly discussed followed by the definition of a (static and dynamic) catastrophe of a dynamical system. A general methodology for the investigation of catastrophes in dynamical systems will follow from the definitions of a catastrophe and a sensitive state.

The classification of sensitive states and the standard forms of dynamical systems describing a system in the vicinity of a specified sensitive state will be given. It is difficult to overestimate the efforts of mathematicians which have led them to classifying the standard forms of codimension one, two and (partly) three. Following the determination of the sensitive state of a given dynamical system, the nature of dynamics of catastrophes of this system may be inferred from knowledge of a suitable standard form.

Classification of catastrophes will be preceded by the centre manifold theorem which is a counterpart to the splitting lemma in elementary catastrophe theory. It will turn out that in the catastrophe theory of dynamical systems such notions of elementary catastrophe theory as the catastrophe manifold, bifurcation set, sensitive state, splitting lemma, codimension, universal unfolding and structural stability are retained.

Subsequently, we will give the method enabling a demonstrative representation of certain dynamical catastrophes, such as the Hopf bifurcation, in the spirit of elementary catastrophe theory. The Hopf bifurcation will be shown to be in principle an elementary catastrophe (with potential). Finally, we shall discuss an application of the above mentioned methods to the examination of catastrophes possible in equations of the reactions with diffusion.

5.2 CLASSIFICATION OF LINEAR SYSTEMS

Models of a diffusionless chemical reaction are described by systems of autonomous ordinary differential equations of first order:

$$dx_i/dt = F_i(x_1, ..., x_n; c_1, ..., c_k), \quad i = 1, ..., n \tag{5.1}$$

where the functions F_i are generally non-linear functions, dependent on n functions of state and k parameters. When equations (5.1) represent a chemical system, state variables x_i are usually concentrations of reagents which must meet the requirement $x_i \geqslant 0$.

Consider the case of a system of two equations, $n = 2$,

$$\dot{x} = P(x, y; \varepsilon) \tag{5.2a}$$

$$\dot{y} = Q(x, y; \varepsilon) \tag{5.2b}$$

where $\varepsilon = (\varepsilon_1, ..., \varepsilon_k)$ stands for control parameters, $x \equiv x_1$, $y \equiv x_2$, $P \equiv F_1$, $Q \equiv F_2$, $\dot{x} = dx/dt$. The solution to this system of equations, $x = x(t)$, $y = y(t)$, describes the motion of the point (x, y), representing the state of the system at time t, in the space of state variables x, y, that is in the so-called phase space (in the case of system (5.2) this is a plane). It follows from Cauchy's theorem that through each point in the phase space may pass just one so-called integral curve $y = y(x)$ or $x = x(y)$ (the set of integral curves of a system is called the phase portrait), whose slope at this point is given by

$$dy/dx = Q(x, y)/P(x, y), \quad \text{or} \quad dx/dy = P(x, y)/Q(x, y) \tag{5.3}$$

except the singular points, $x = \bar{x}$, $y = \bar{y}$, at which vanish simultaneously the right-hand sides of equation (5.3)

$$P(\bar{x}, \bar{y}) = 0, \quad \text{or} \quad Q(\bar{x}, \bar{y}) = 0 \tag{5.4a}$$

At such points several integral curves of equation (5.3) may intersect. The singular points $(\bar{x}, \bar{y})$ also have a physical interpretation. It follows from equations (5.2a), (5.2b) that $d\bar{x}/dt = 0$, $d\bar{y}/dt = 0$, i.e. the points $(\bar{x}, \bar{y})$ are stationary points of the system (5.2). When equations (5.2) or (5.1) describe chemical systems, stationary points (sometimes also called rest points) must satisfy the additional condition

$$\bar{x} \geqslant 0, \quad \bar{y} \geqslant 0 \tag{5.4b}$$

if the variables x, y represent concentrations of X, Y.

Properties of stationary points, for example their number and stability (the stationary point $(\bar{x}, \bar{y})$ is stable if the points $x(t)$, $y(t)$, close to the point $\bar{x}$, $\bar{y}$, approach it when $t \to \infty$) may be examined by linearizing the system (5.2) in the neighbourhood of its stationary points. To investigate the stability of the stationary point $(\bar{x}, \bar{y})$ we make a substitution in (5.2)

$$x = \bar{x} + \xi, \quad y = \bar{y} + \eta \tag{5.5}$$

and, assuming that ξ, η are small, we neglect all powers of ξ, η of order higher than one, obtaining the linear system of equations

$$\dot{\xi} = a_{11}(\varepsilon)\xi + a_{12}(\varepsilon)\eta \tag{5.6a}$$

$$\dot{\eta} = a_{21}(\varepsilon)\xi + a_{22}(\varepsilon)\eta \tag{5.6b}$$

$$a_{ij}(\varepsilon) = [\partial F_i/\partial x_j]_{(\bar{x}_1, \bar{x}_2)}, \quad x_{1,2} \equiv x, y, \quad F_{1,2} \equiv P, Q \tag{5.6c}$$

A stationary point of the linearized system (5.6) is the origin of the coordinate system $(\bar{\xi}, \bar{\eta}) = (0, 0)$. The matrix $F_{ij} = \partial F_i/\partial x_j$ will be called the Jacobian matrix whereas the matrix $a_{ij} = F_{ij}(\bar{x}_1, \bar{x}_2)$ will be called the stability matrix of the system (5.2).

The linear system (5.6) may be readily solved by the replacement

$$\xi = \xi_0 e^{\lambda t}, \quad \eta = \eta_0 e^{\lambda t} \tag{5.7}$$

(ξ_0, η_0 are time-independent) which allows to reduce the problem (5.2) to an algebraic equation (this is the so-called eigenproblem):

$$a_{11}\xi_0 + a_{12}\eta_0 = \lambda \xi_0 \tag{5.8a}$$

$$a_{21}\xi_0 + a_{22}\eta_0 = \lambda \eta_0 \tag{5.8b}$$

Apparently, the condition for existence of non-zero solutions to the system of equations (5.8) is vanishing of the determinant

$$\begin{vmatrix} a_{11} - \lambda & a_{12} \\ a_{21} & a_{22} - \lambda \end{vmatrix} = \lambda^2 - (a_{11} + a_{22})\lambda + (a_{11}a_{22} - a_{12}a_{21}) = 0 \tag{5.9}$$

which yields two values of the parameter λ, λ_1, λ_2, obviously depending in view of equation (5.6c) on the parameter ε.

Thus, the solutions to the linearized system are of the form

$$x(t) = \bar{x} + c_1 \exp(\lambda_1 t) + c_2 \exp(\lambda_2 t) \tag{5.10a}$$

$$y(t) = \bar{y} + d_1 \exp(\lambda_1 t) + d_2 \exp(\lambda_2 t) \tag{5.10b}$$

and their evolution with time depends on the λ_1, λ_2 values (eigenvalues). The following cases characterizing a type of the singular point $\bar{x}$, $\bar{y}$ and a stability of solution (5.10) to the linearized system (5.8) can be distinguished:

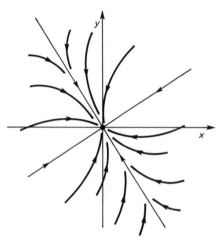

Fig. 58. Phase trajectories for the system (5.2) in the vicinity of the stationary point of a stable node type.

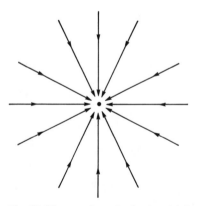

Fig. 59. Phase trajectories in the vicinity of the stationary point of a stable starlike node type.

(a1) $\lambda_1 < 0$, $\lambda_2 < 0$ $(\lambda_1 = \lambda_2)$. In this case solutions (5.10) have the following property: $x(+\infty) = \bar{x}$, $y(+\infty) = \bar{y}$, that is the system when deflected from the stationary state $\bar{x}$, $\bar{y}$ returns to it. Such a singular point is called a stable node. Figure 58 illustrates the shape of trajectories in the phase space for various positions of the starting point at the time t_0, $x(t_0)$, $y(t_0)$.

(a2) $\lambda_1 = \lambda_2 = \lambda$, $\lambda < 0$, the matrix of coefficients a_{ij} is proportional to the identity matrix E_{ij}, $a_{ij} \sim E_{ij}$. All the rays converging to the stationary point: $x(+\infty) = \bar{x}$, $y(+\infty) = \bar{y}$, constitute trajectories, see Fig. 59. Such a stationary point is called a starlike (stable) node. The system is structurally unstable: an arbitrarily small perturbation may result in the change of a type of the system to (a1) or (d).

(a3) $\lambda_1 = \lambda_2 = \lambda$, $a_{ij} \sim E_{ij}$. The trajectories enter the stationary point $x(+\infty) = \bar{x}$, $y(+\infty) = \bar{y}$, see Fig. 60. This is a degenerate stable node. Such a system is structurally unstable: an arbitrarily small perturbation may change a type of the system to (a1) or (d).

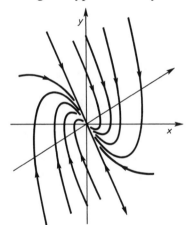

Fig. 60. Phase trajectories in the vicinity of the stationary point of a degenerate stable node type.

(b1) $\lambda_1 > 0$, $\lambda_2 > 0$ $(\lambda_1 = \lambda_2)$. In this case $x(-\infty) = \bar{x}$, $y(-\infty) = \bar{y}$. Such a point is called an unstable node. A system deflected from the stationary state $\bar{x}$, $\bar{y}$ moves away from it. Typical trajectories in the phase space are shown in Fig. 61.

(b2) $\lambda_1 = \lambda_2 = \lambda$, $\lambda > 0$, $a_{ij} \sim E_{ij}$. The trajectories consist of all rays diverging from the stationary point: $x(-\infty) = \bar{x}$, $y(-\infty) = \bar{y}$, see Fig. 62. Such a stationary point is called a starlike (unstable) node. The system is structurally unstable: an arbitrarily small perturbation may result in the change of a type of the system to (b1) or (e).

(b3) $\lambda_1 = \lambda_2 = \lambda > 0$, $a_{ij} \sim E_{ij}$. The trajectories enter the stationary point $x(-\infty) = \bar{x}$, $y(-\infty) = \bar{y}$, see Fig. 63. This is a degenerate unstable node. The system is structurally unstable: an arbitrarily small perturbation may change the system type to (b1) or (e).

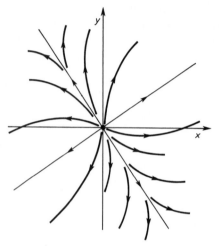

Fig. 61. Phase trajectories in the vicinity of the stationary point of an unstable node type.

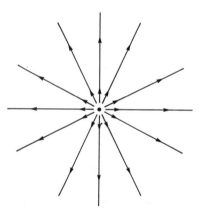

Fig. 62. Phase trajectories in the vicinity of the stationary point of an unstable starlike node type.

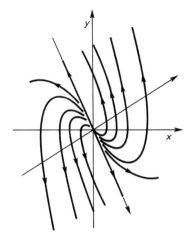

Fig. 63. Phase trajectories in the vicinity of the stationary point of a degenerate unstable node type.

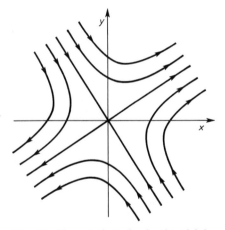

Fig. 64. Phase trajectories in the vicinity of the stationary point of a saddle type.

(c) $\lambda_1 > 0$, $\lambda_2 < 0$ (or $\lambda_1 < 0$, $\lambda_2 > 0$). The system deflected from the stationary state moves away from it with a probability equal to one. Such a point is called a saddle. Phase trajectories are shown in Fig. 64.

(d) $\lambda_{1,2} = \alpha \pm i\beta$, $\alpha < 0$ ($i^2 = -1$). In this case $x(+\infty) = \bar{x}$, $y(+\infty) = \bar{y}$. The system deflected from the state $\bar{x}$, $\bar{y}$ will oscillate with a decreasing amplitude, returning to the stationary point, see Fig. 65. Such a point is called a stable focus.

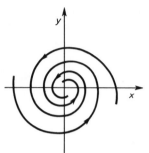

Fig. 65. Phase trajectories in the vicinity of the stationary point of a stable focus type.

(e) $\lambda_{1,2} = \alpha \pm i\beta$, $\alpha > 0$. In this case $x(-\infty) = \bar{x}$, $y(-\infty) = \bar{y}$. The system deflected from the state $\bar{x}$, $\bar{y}$ will oscillate with an increasing amplitude, moving away from the stationary point, see Fig. 66. Such a point is called an unstable focus.

(f) $\lambda_{1,2} = \pm i\beta$, $(\alpha = 0)$. In this case $x(-\infty) = \bar{x}$, $y(-\infty) = \bar{y}$. The system deflected from the state $\bar{x}$, $\bar{y}$ will unceasingly oscillate around the stationary point, see Fig. 67. Such a point is called a centre. The system is

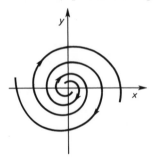

Fig. 66. Phase trajectories in the vicinity of the stationary point of an unstable focus type.

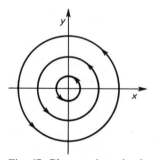

Fig. 67. Phase trajectories in the vicinity of the stationary point of a centre type.

structurally unstable: an arbitrarily small perturbation of the real part of the eigenvalues $\lambda_{1,2}$ converts a centre into a stable focus (d) or unstable focus (e), cf. Figs. 65, 66.

(g) $\lambda_1 = 0$, $\lambda_2 < 0$. Phase trajectories are parallel straight lines (Fig. 68). The system is structurally unstable. An arbitrarily small perturbation may

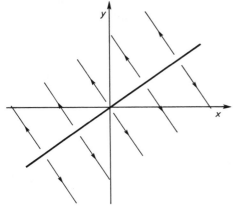

Fig. 68. Phase trajectories in the vicinity of the stationary point for the case $\lambda_1 = 0$, $\lambda_2 < 0$.

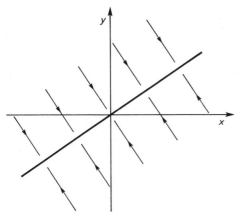

Fig. 69. Phase trajectories in the vicinity of the stationary point for the case $\lambda_1 = 0$, $\lambda_2 > 0$.

convert it into a stable node (a1) or into a saddle (c), cf. Figs. 58, 64.

(h) $\lambda_1 = 0$, $\lambda_2 > 0$. Phase trajectories are parallel straight lines (Fig. 69). The system is structurally unstable. An arbitrarily small perturbation may convert it into an unstable node (b1) or a saddle (c), cf. Figs. 61, 64.

(i1) $\lambda_1 = \lambda_2 = 0$, $a_{ij} = 0$. Each point of the phase space is an equilibrium position (this case is related to cases (a2) and (b2). The system is structurally unstable. An arbitrarily small disturbance may convert it, for example, into the system with a stationary point of a starlike node (stable or unstable) type.

(i2) $\lambda_1 = \lambda_2 = 0$, $a_{ij} \neq 0$. The case is related to cases (a3) and (b3). Phase trajectories are shown in Fig. 70. The system is structurally unstable. An arbitrarily small perturbation may convert it, for example, into a stable (a3) or unstable (b3) degenerate node, cf. Figs. 60, 62.

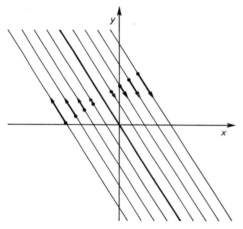

Fig. 70. Phase trajectories in the vicinity of the stationary point for the case $\lambda_1 = \lambda_2 = 0$, $a_{ij} = 0$.

The stability of stationary points in cases (a1), (a2), (d) is proven in Appendix 45 using the Lyapunov function method.

The regions of phase space, filled with trajectories of similar asymptotic behaviour, are separated by the sets (for $n = 2$ these are straight lines) called separatrices. The directions along which separatrices reach a stationary point are called the main directions (axes). The method of determination of separatrices is given in Appendix A2.6.

5.3 NON-LINEAR DYNAMICAL SYSTEMS

5.3.1 The Grobman-Hartman theorem

The behaviour of trajectories of the system (5.2) in the neighbourhood of the stationary point $\bar{x}$, $\bar{y}$ is generally determined by the properties of the linearized system (5.6). It can be shown that when the following conditions are satisfied:

(1) the functions P, Q have continuous derivatives up to and including second order,

(2) $\det [a_{ij}] \neq 0$,

(3) the eigenvalues $\lambda_{1,2}$ are not purely imaginary,

then the stationary point $\bar{x}$, $\bar{y}$ of the system (5.2) is of the same type as the stationary point $(0, 0)$ of the system (5.6). Trajectories of the system (5.2) in the vicinity of the stationary point are deformed images of trajectories of the system (5.6), the deformation increasing with the distance from the point $(\bar{x}, \bar{y})$. This theorem is called the Hartman–Grobman theorem. The stationary points fulfilling conditions (1)–(3) may be regarded as corresponding to nondegenerate critical points of elementary catastrophe theory.

The Grobman–Hartman theorem provides a criterion permitting to establish the equivalence (nonequivalence) of two non-linear autonomous systems when conditions (1)–(3) are satisfied. In this way, a counterpart to the relation of equivalence of potential functions of elementary catastrophe theory (see Chapter 2) is obtained in the area of autonomous systems.

The analysis of linearized sytem thus allows, when conditions (1)–(3) are met, us to find the shape of phase trajectories in the vicinity of stationary (singular) points. A further, more thorough examination must answer the question what happens to trajectories escaping from the neighbourhood of an unstable stationary point (unstable node, saddle, unstable focus). In a case of non-linear systems such trajectories do not have to escape to infinity. The behaviour of trajectories nearby an unstable stationary point will be examined in further subchapters using the catastrophe theory methods.

When conditions (1)–(3) are not fulfilled, the Grobman–Hartman theorem is not valid. As will be shown later, then we have to deal with the sensitive state of a dynamical system (this corresponds to a degenerate critical point in elementary catastrophe theory). A generalization of the Grobman–Hartman theorem, the centre manifold theorem which may be regarded as a counterpart to the splitting lemma of elementary catastrophe theory, has been found to be very convenient in that case.

When condition (3) is not fulfilled, we deal in the case of system (5.2) with a structurally unstable centre. In real physical systems this kind of a stationary point is not encountered. As already mentioned above, an arbitrarily small perturbation of the centre may lead to its conversion into a focus; thus, linearization of the system (5.2) does not allow us to draw conclusions on the shape of trajectories near the stationary point of a non-linear system.

When condition (2) is not met, we have to deal with a stationary point of higher order, for which the linearization method also fails. It appears that there exists a great variety of types of stationary points of higher order.

5.3.2 Limit sets

To characterize the limit behaviour of trajectories at $t \to \pm \infty$, the notion of a limit set has been introduced. If a sequence $x(t_1)$, $x(t_2)$, ... for $t_i \to \infty$ (or for $t_i \to -\infty$) is convergent to a point p, then the point p is called the ω-limit (α-limit) point of a given trajectory. A set of all limit points of a specified trajectory is called the limit set. Such an (attracting) limit set is also referred to as the attractor.

The notions of a limit point and a limit set will be first exemplified by the linear systems considered in Section 5.1. In the case of a stable node (a1), (a2), (a3) and a stable focus (d) the limit set (attractor) consists of one point, the stationary point, which is approached by all trajectories.

In the non-linear systems (5.2), a second type of attractor — a closed curve (limit cycle) is also possible. For example, the system of van der Pol equations (representing oscillations of current in electrical circuits and oscillations of concentrations, or more precisely the differences between the concentrations and their stationary values, in chemical systems)

$$\dot{x} = y \tag{5.11a}$$

$$\dot{y} = -x + \varepsilon y (1 - x^2) \tag{5.11b}$$

where ε is a parameter, has one stationary point $(0, 0)$. Linearizing the system (5.12) nearby the point $(0, 0)$ we obtain a linear system and the respective eigenvalues, see equation (5.9)

$$\dot{\xi} = \eta \tag{5.12a}$$

$$\dot{\eta} = -\xi + \varepsilon y \tag{5.12b}$$

$$\lambda_{1,2} = 1/2 \left[\varepsilon \pm (\varepsilon^2 - 4)^{1/2} \right] \tag{5.13}$$

that is, the stationary point (0, 0) for $0 < \varepsilon < 2$ is an unstable focus (case (e)). A full analysis of the van der Pol system, see Section 5.6, leads to a conclusion that trajectories of the system, both those escaping from an unstable focus and those coming out of infinity, wind on a certain closed curve: we then say that the limit set is an attractor − a limit cycle. On the other hand, when $-2 < \varepsilon < 0$, the stationary point is a stable focus (case (d)). Both the cases are illustrated in Fig. 71.

(a)   (b)

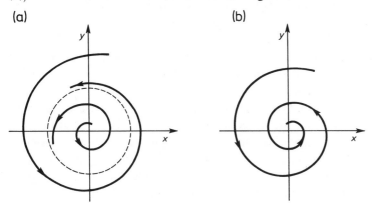

Fig. 71. The van der Pol system: (a) $0 < \varepsilon < 2$, (b) $-2 < \varepsilon < 0$.

It may be demonstrated (the Poincaré–Bendixon theorem, see Appendix A2) that for the autonomous system in two variables (5.2) a limit set can only be a point or a closed curve. Linear systems, gradient and Hamiltonian systems, will be shown in Appendix A1 to be uncapable of having a limit cycle.

In a case of autonomous systems depending on three or more variables there exist more types of limit sets which, in some cases, may be extraordinarily complex. We shall discuss below an important case of a very complex limit set, which may occur in a dynamical system defined in the three-dimensional phase space. The following system of equations, known as the Lorenz system, in which σ, r, b are certain control parameters,

$$\dot{x} = -\sigma x + \sigma y \qquad\qquad\qquad\qquad (5.14a)$$

$$\dot{y} = rx - y - xz \qquad\qquad\qquad\qquad (5.14b)$$

$$\dot{z} = -bz + xy \qquad\qquad\qquad\qquad (5.14c)$$

(in the case of heat convection σ is the Prandtl number, r is the normalized Rayleigh number, $b = 8/3$), describes, among other phenomena, heat convection, laser emission, changes in the position of geomagnetic poles.

Stationary points of the system (5.14), i.e. those satisfying the equations $\bar{x} = 0$, $\bar{y} = 0$, $\bar{z} = 0$, are given by

$$(\bar{x}_1, \bar{y}_1, \bar{z}_1) = (0, 0, 0), \quad \text{for all } r \text{ values} \tag{5.15}$$

$$(\bar{x}_2, \bar{y}_2, \bar{z}_2) = (\alpha, \alpha, \gamma), \quad (\bar{x}_3, \bar{y}_3, \bar{z}_3) = (-\alpha, -\alpha, \gamma) \tag{5.16}$$

where $\gamma \equiv r - 1$, $\alpha \equiv (b\gamma)^{1/2}$, for $r > 1$.

When the control parameters σ, b have fixed values, for example $\sigma = 10$, $b = 8/3$, and the parameter r increases from the value $r = 0$, the nature of phase trajectories of the system (5.14) changes qualitatively. We shall describe properties of the Lorenz system only for the values of parameter r larger than $r \cong 24.74$ and smaller than a certain value r_∞ (Lorenz in his work has studied the case $\sigma = 10$, $b = 8/3$, $r = 28 < r_\infty$). When $24.74 < < r < r_\infty$, all three stationary points are unstable. However, trajectories do not escape to infinity — all the trajectories are attracted by a certain region of the phase space (attractor), containing the stationary points and which is approximately a two-dimensional surface.

On the other hand, it follows from the Liouville theorem (Appendix A2.1) that the attractor volume must be zero. Indeed, the divergence of the vector field $\mathbf{F}$ constructed from the right-hand sides of the Lorenz equation, $\mathbf{F} = (-\sigma x + \sigma y, \ rx - y - xz, \ -bz + xy)$, is equal to

$$\text{div } \mathbf{F} = \partial(-\sigma x + \sigma y)/\partial x + \partial(rx - y - xz)/\partial y + \partial(-bz + xy)/\partial z$$
$$= -(1 + \sigma + b) \tag{5.17}$$

and hence it is a negative quantity. Using now the Liouville theorem we come to the conclusion that the phase trajectories attracted by the attractor and approaching it more and more closely for $t \to \infty$, are present in the region of an increasingly smaller volume. From that follows directly the inference about zero volume of the attracting limit set (attractor).

All the trajectories being present in the field of action of the attractor are unstable and behave chaotically, see Fig. 72.

On the other hand, it follows from the Poincaré–Bendixon theorem (see Appendix A2.2) that an attractor cannot have a dimension equal to two (plane), one (line) or zero (point), since a chaotic nature of the limit set is then impossible (only stationary points and limit sets are then possible). Hence, a conclusion follows, confirmed by other methods, that the attractor of the system of equations (5.14) has a fractional dimension (more exactly, $2 + D$, where D is a small positive number), see Appendix A2.7.

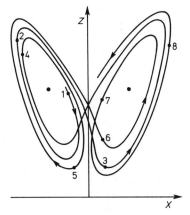

Fig. 72. Chaotic trajectories for the Lorenz system.

This attractor is called the Lorenz attractor. The Lorenz attractor falls into a class of so-called strange attractors, corresponding to the quasi--stochastic behaviour of a system. Dynamics of this type has been found to occur in the systems of chemical kinetics equations (see Section 6.3.3).

5.4 FUNDAMENTALS OF CATASTROPHE THEORY OF DYNAMICAL SYSTEMS

5.4.1 Definition of a catastrophe of a dynamical system

We say that $c = c_0$ is the point of occurrence of a catastrophe for the system (5.1) when a continuous change in control parameter values is accompanied by the variation in qualitative character of the phase portrait at c_0.

Thus, a catastrophe will take place in the dynamical system (5.1) if:

(1) the number of stationary points of the system $(\bar{x}_1, ..., \bar{x}_n)$ with non-negative coordinates $\bar{x}_1 \geqslant 0, ..., \bar{x}_n \geqslant 0$ changes;

(2) the type of at least one stationary point changes.

When a catastrophe occurs in a dynamical system, two cases are possible. In the first case, a static catastrophe, the only stable states of the system are stationary points; in the case of dynamical catastrophes, stable non-stationary solutions, for example limit cycles, appear.

5.4.2 Gradient systems

A gradient system can be written in the following form

$$\dot{x}_i = -\partial V/\partial x_i, \quad i = 1, ..., n \tag{5.18}$$

where $V(x_1, ..., x_n; c_1, ..., c_k)$ is a potential function dependent on n state variables and k parameters.

Stationary states of a gradient system fulfil (equivalent) equations

$$\dot{x}_i = 0 \tag{5.19a}$$

$$\nabla_x V = 0 \tag{5.19b}$$

States of the system may be divided into three groups: (a) nonstationary state, $\nabla_x V \neq 0$; (b) nondegenerate stationary state, $\nabla_x V = 0$, $\det(V_{ij}) \neq 0$ ($V_{ij} \equiv \partial^2 V/\partial x_i \partial x_j$); (c) degenerate stationary state $\nabla_x V = 0$, $\det(V_{ij}) = 0$; analogously to the classification used in elementary catastrophe theory.

Stationary states of a gradient system may be said to lie on the catastrophe surface M and degenerate stationary states to be placed on the singularity set Σ (see Chapter 2).

The fundamental stages of investigation of a control parameter-dependent gradient system nearby nondegenerate critical points will now be described:

(1) for fixed values of control parameters all stationary states are found from the requirement $\nabla_x V = 0$, followed by the determination of their type and the main axes;

(2) a phase portrait of the system is then constructed.

The properties of a nondegenerate stationary point, and its stability, derive from the properties of nondegenerate critical points of the potential. Note that the classification of stationary points described in Section 5.1 applies directly to a nondegenerate stationary point of a gradient system. Hence, in the case of gradient systems, the requirement of lack of degeneracy of a critical point constitutes the criterion of applicability of this classification.

In the case of a degenerate critical point, $\det(V_{ij}) = 0$ and the linear approximation described in Section 5.1 fails. The behaviour of the system may be examined by expanding in a Taylor series a potential function about the degenerate stationary point $\bar{x}$, in accordance with principles of expansion in a Taylor series of a function in the vicinity of a degenerate critical point, and solving the system of equations obtained in this manner.

However, it is frequently much more convenient to proceed in the following way: introduce a small perturbation to a potential function in such a way that the perturbed system has isolated stationary points and examine the phase portrait nearby the stationary points. Such an approach, involving the replacement of a degenerate stationary point by isolated, very closely located nondegenerate critical points, is called morsification.

5.4.3 Autonomous non-gradient systems

The necessary and sufficient condition for the autonomous system

$$\dot{x}_i = F_i(\mathbf{x}; \mathbf{c}), \qquad i = 1, ..., n \tag{5.20}$$

to be written in the gradient form (5.18) is that the requirements

$$\partial F_i / \partial x_j - \partial F_j / \partial x_i = 0, \quad i, j = 1, ..., n \tag{5.21}$$

be satisfied.

In the analysis of autonomous non-gradient systems the methods of gradient system theory have proved useful. The notions such as a stationary (critical) point, degenerate stationary point, structural stability (instability), morsification, phase portrait can be directly transferred to autonomous systems. A qualitative description of dynamical autonomous systems is constructed analogously with the description of gradient systems.

A general program of examination of autonomous dynamical systems may be formulated, in the spirit of elementary catastrophe theory, as follows:

(1) finding the stationary states of a dynamical system;
(2) determination of the bifurcation set in R^n, associated with the stability matrix F;
(3) construction of the most general deformation of dynamical systems, described by points of each of the components of the bifurcation set;
(4) determination of geometrical properties of every such deformation.

Further analysis of non-gradient systems will be carried out for the case of a system of two equations ($n = 2$):

$$\dot{x} = F_1(x, y) \tag{5.22a}$$

$$\dot{y} = F_2(x, y) \tag{5.22b}$$

Stationary states of a non-gradient system fulfil (equivalent) equations

$$\dot{x} = 0, \quad \dot{y} = 0 \tag{5.23a}$$

$$F_1(\bar{x}, \bar{y}) = 0, \quad F_2(\bar{x}, \bar{y}) = 0 \tag{5.23b}$$

The conditions (5.23) are similar to the requirements (5.19) for gradient systems.

States of a system may be classified into three groups: (a) the state beyond a stationary point, $\mathbf{F}(x, y) \neq 0$; (b) nondegenerate stationary state, $\mathbf{F}(x, y) = 0$, $\det(\partial F_i/\partial x_j) \neq 0$; (c) degenerate stationary state, $\mathbf{F}(x, y) = 0$, $\det(\partial F_i/\partial x_j) = 0$; analogously with the classification of gradient systems.

The stationary states of the system may be said to lie on the catastrophe surface M; degenerate stationary states are located on the singularity set Σ (see Chapter 2).

The stability of stationary states may be investigated by linearizing the system (5.22) nearby an equilibrium state, see Section 5.1, equations (5.6).

The above classification of states of a system into three groups can be done in an equivalent way be using the properties of the stability matrix:
(a) non-stationary state, $\dot{\mathbf{x}} \neq 0$ or $\mathbf{F} \neq 0$;
(b) nondegenerate stationary state, $\mathbf{F} = 0$, $\det(a_{ij}) \neq 0$;
(c) degenerate stationary state, $\mathbf{F} = 0$, $\det(a_{ij}) = 0$;
by analogy to the classification of gradient systems.

If eigenvalues are real and different, then a given dynamical system is locally (in the vicinity of a stationary point) equivalent to a certain structurally stable gradient system (this is an unstable node when $\lambda_1 > 0$, $\lambda_2 > 0$; a saddle when $\lambda_1 > 0$, $\lambda_2 < 0$ or $\lambda_1 < 0$, $\lambda_2 > 0$; a stable node when $\lambda_1 < 0$, $\lambda_2 > 0$). In the remaining cases a dynamical system is not locally equivalent to a gradient system.

5.4.4 Sensitive states of autonomous systems

The analysis of properties of gradient systems carried out in terms of elementary catastrophe theory (examination of critical points of the potential V) and of nongradient systems by means of singularity theory (examination of singularities of the vector function $\mathbf{F}$) provides an incentive to investigate the relation between possible catastrophes and the eigenvalues of the stability matrix.

Note that vanishing of the determinant $\det(\partial^2 V/\partial x_i \partial x_j)$ or $\det(\partial F_i/\partial x_j)$, corresponding to the occurrence of a catastrophe, implies vanishing of the stability matrix determinant:

$$\det(\partial F_i/\partial x_j)(0) = 0 \leftrightarrow \det(a_{ij}) = 0 \qquad (5.24)$$

Hence, let us characterize more exactly, in terms of eigenvalues of the stability matrix, the sensitive state appearing in such a case.

Consider the system of two equations (5.2), having the stationary point $(0, 0)$ for all values of the control parameters ε. The point $\varepsilon = 0$ is said to be the point of occurrence of a catastrophe (also called a bifurcation) for the system (5.25), if the qualitative nature of a phase portrait changes in the vicinity of the stationary point $(0, 0)$ on crossing by control parameters the value $\varepsilon = 0$.

It follows from equation (5.9) that the requirement $\det(a_{ij}) = 0$ is equivalent to vanishing of at least one eigenvalue, for example λ_1:

$$\det(a_{ij}) = 0 \leftrightarrow \lambda_1 = 0, \qquad \lambda_2 = \lambda \qquad (5.25)$$

Such a case is undoubtedly the sensitive state, since an arbitrarily small change in the control parameter ε_k may result in the appearance of a negative or positive λ_1 value and in a qualitative change in the phase portrait, i.e. in a catastrophe. In other words, the case (5.25) divides qualitatively different states.

The above observation may be employed for finding the remaining sensitive states of the system (5.2) on the basis of knowledge of its stability matrix a_{ij} and the classification of stationary states given in Section 5.1. The sensitive states are those corresponding to such values of the control parameters ε that the eigenvalues of the stability matrix $a_{ij}(\varepsilon)$ are of the following form:

(I) $\text{Re}(\lambda_1) = 0$ $\hspace{6cm}$ (5.26a)

(IA) $\lambda_1 = 0, \quad \lambda_2 = \lambda$ $\hspace{4.5cm}$ (5.26a1)

(IB) $\lambda_{1,2} = \pm i\beta$ $\hspace{5.5cm}$ (5.26a2)

(II) $\lambda_1 = \lambda_2 = \lambda$ $\hspace{5cm}$ (5.26b)

where λ is real (and, in a special case, λ may be zero).

Note that the above classification in an evident way extends the classification (5.24) of elementary catastrophe theory and singularity theory: for $n = 2$ the criterion (5.26) employs the trace $(a_{11} + a_{22})$ and the determinant $(a_{11}a_{22} - a_{12}a_{21})$ of the stability matrix, whereas the criterion (5.24) takes into account only the determinant of the matrix.

For λ different from zero, the sensitive states (5.26) separate the following states

(IA) $\lambda_1 = -\varepsilon_1, \quad \lambda_2 = (\lambda + \varepsilon_2)$

        $\lambda_1 = +\varepsilon_1, \quad \lambda_2 = (\lambda + \varepsilon_2)$ (5.27a1)

(IB) $\lambda_1 = -\varepsilon_1 - i(\beta + \varepsilon_2), \quad \lambda_2 = -\varepsilon_1 + i(\beta + \varepsilon_2)$

        $\lambda_1 = +\varepsilon_1 - i(\beta + \varepsilon_2), \quad \lambda_2 = +\varepsilon_1 + i(\beta + \varepsilon_2)$ (5.27a2)

(II) $\lambda_1 = \lambda + \varepsilon_1, \quad \lambda_2 = \lambda + \varepsilon_2$

      $\lambda_1 = (\lambda + \varepsilon_1) - i\varepsilon_2, \quad \lambda_2 = (\lambda + \varepsilon_1) + i\varepsilon_2$ (5.27b)

where $|\varepsilon_1|, |\varepsilon_2| \ll 1$. When $\lambda < 0$, $\varepsilon_1 > 0$, the phase portrait of a sensitive system (structurally unstable) and the systems divided by it are shown in figures: for (IA) in Figs. 58, 68, 64; for (IB) in Figs. 65, 66, 67; for (II) in Figs. 58, 60, 65.

As we can see, in all the three cases a qualitative change in the phase portrait takes place. However, there is a fundamental difference between cases (IA), (IB) and (II): in cases (IA), (IB) the change in the phase portrait nearby a stationary point occurs while in case (II) the neighbourhood of the stationary point occurs while in case (II) the neighbourhood of the stationary point does not vary, but a global character of the phase portrait is altered. In other words, the catastrophe appearing in case (II) is finer and more difficult to examine.

The above result is related to the crucial Grobman–Hartman theorem, which provides the necessary and sufficient condition for the change in a phase portrait in the vicinity of a stationary point (i.e. pertains to a generalization of cases (IA), (IB), see Section 5.3.1. It follows from the Grobman–Hartman theorem that in dynamical systems consisting of a larger number of equations ($n > 2$) a catastrophe of the change in a phase portrait nearby a stationary point may occur only if the system upon varying a control parameter crosses the sensitive state such that the eigenvalues of the stability matrix $a_{ij}(\varepsilon)$ are of the following form:

(I) $\mathrm{Re}(\lambda_1) = \ldots = \mathrm{Re}(\lambda_p) = 0$ (5.28)

5.4.5 The centre manifold theorem

When we deal with a dynamical system in which a sensitive state occurs (assumptions of the Grobman–Hartman theorem are not fulfilled), it may turn out that the sensitive state is associated only with a part of state variables. The variables related to the sensitive state may then be separated and the catastrophes occurring in a system dependent on a smaller number of state variables examined.

The circumstances resembles very much the splitting lemma in elementary catastrophe theory. Recall that a function of many variables, having a degenerate critical point (sensitive state), may be separated into two parts: the part dependent on a number of state variables, having no degenerate critical point, and the part depending on the remaining state variables, having a degenerate critical point.

A counterpart to the splitting lemma of elementary catastrophe theory is the centre manifold theorem (also called a neutral manifold), generalizing the Grobman–Hartman theorem to the case of occurrence of sensitive states (5.28). The centre manifold theorem allows us to establish an equivalence (nonequivalence) of two autonomous systems. In this sense it is also a generalization to the case of autonomous systems of the equivalence relationship introduced in Chapter 2 for potential functions.

Consider the following system of equations

$$\dot{x}_i = \sum_k^n B_{ik}(\mathbf{c})x_k + f_i(\mathbf{x}, \mathbf{y}; \mathbf{c}) \tag{5.29a}$$

$$\dot{y}_j = \sum_l^m C_{jl}(\mathbf{c})y_l + g_j(\mathbf{x}, \mathbf{y}; \mathbf{c}) \tag{5.29b}$$

where B_{ik} is the $n \times n$ matrix, C_{jl} is the $m \times m$ matrix, $\mathbf{x} \in R^n, \mathbf{y} \in R^m$. The vector $\mathbf{c} \in R^k$ corresponds to control parameters.

Let for $\mathbf{c} = \mathbf{c}_0$ the functions f_i, g_j vanish together with their first derivatives at $(\mathbf{x}, \mathbf{y}) = (0, 0)$ (the origin of the coordinate system), that is, not contain linear terms. Then the system (5.29) has the stationary point $(0, 0)$. Linearization of the system (5.29) in the neighbourhood of the point $(0, 0)$ yields the linear system

$$\dot{\xi}_i = \sum_k^n B_{ik}(\mathbf{c}_0)\xi_k \tag{5.30a}$$

$$\dot{\eta}_j = \sum_l^m C_{jl}(\mathbf{c}_o)\eta_l \tag{5.30b}$$

Substitution of the type (5.7) in equations (5.30)

$$\xi_i = \xi_{0,i}e^{\lambda t}, \qquad \eta_j = \eta_{0,j}e^{\mu t} \tag{5.31}$$

leads to two independent eigenequations:

$$\sum_k^n (B_{ik} - \lambda\delta_{ik})\xi_{0,k} = 0 \tag{5.32a}$$

$$\sum_l^m (C_{jl} - \mu\delta_{jl})\eta_{0,l} = 0 \tag{5.32b}$$

where δ_{ik} is equal to 1 when $i = k$ and to 0 in the remaining cases.

Assume now that the eigenvalues of the matrices B_{ik} and C_{jl}, satisfying the equations

$$\det|B_{ik} - \lambda\delta_{ik}| = 0 \tag{5.33a}$$

$$\det|C_{jl} - \mu\delta_{jl}| = 0 \tag{5.33b}$$

have the following properties:

$$\mathrm{Re}(\lambda_1) = ... = \mathrm{Re}(\lambda_n) = 0 \tag{5.34a}$$

$$\mathrm{Re}(\mu_1) < 0, ..., \mathrm{Re}(\mu_m) < 0 \tag{5.34b}$$

Let us briefly examine properties of the system (5.29). Firstly, the stationary state (0, 0) is a sensitive state in view of equations (5.34a). Secondly, to simplify notation we assumed that there are no unstable solutions for which $\mathrm{Re}(\lambda) > 0$. Furthermore, it is apparent that the sensitive state is associated only with the variable $\mathbf{x}$.

The centre manifold method permits us to split the system (5.29) into the subsystem having a sensitive state and the subsystem fulfilling assumptions of the Grobman–Hartman theorem. Note that in the linearized system (5.30) the separation took place automatically as a result of linearity. The phase portrait of the linearized system is of the form shown in Fig. 73.

The $\mathbf{x}$-axis is the so-called central surface (also called neutral or slow surface), the $\mathbf{y}$-axis is the so-called stable surface. As a result of fast processes, trajectories of the system approach the $\mathbf{x}$-axis parallel to the $\mathbf{y}$-axis.

The centre manifold theorem deals with the modification of this pattern in the non-linear system (5.29). It follows from the centre manifold

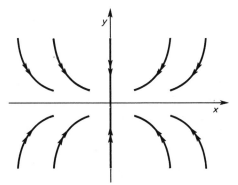

Fig. 73. Phase trajectories and the central surface for the linear system (5.30).

theorem that in the non-linear system (5.29) fast processes reduce trajectories of the system to the centre manifold (slow surface) W_c tangent to the **x**-axis whereas directions of the trajectories approaching W_c are parallel to the surface W_s — the stable manifold. The surface W_s, see Fig. 74, undergoes a deformation compared with the linear variant (5.30), Fig. 73, but remains tangent to the **y**-axis.

The geometrical meaning of the centre manifold theorem is expressed by the graphs of phase trajectories, Fig. 74 (for the non-linear system (5.29) and Fig. 73 (for the linearized system (5.30)).

As the centre manifold is tangent to the **x**-axis, it may be represented in the form

$$W_c = \{(\mathbf{x}, \mathbf{y}): \mathbf{y} = \mathbf{h}(\mathbf{x}) = 0, \quad (\partial h_j / \partial x_i) = 0\} \tag{5.35}$$

where the function **h** is defined in the vicinity of a stationary point.

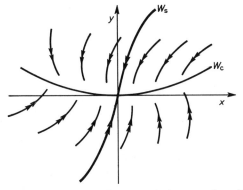

Fig. 74. Phase trajectories and the central surface for the system (5.29).

We may now replace the system (5.29) by the reduced system

$$\dot{x}_i = \sum_k^n B_{ik}(\mathbf{c}) x_k + f_i[\mathbf{x}, \mathbf{h}(\mathbf{x}); \mathbf{c}] \tag{5.36}$$

which constitutes a good approximation of the system (5.29) near the stationary point (0, 0). Linearization of the system (5.36) in the vicinity of the point (0) yields equations (5.30a) and, in view of equations (5.34a), the system (5.36) contains all sensitive states of the system (5.29).

The method of determination of the function $\mathbf{h}(\mathbf{x})$ remains to be discussed. Substituting for $\mathbf{y} = \mathbf{h}(\mathbf{x})$ in (5.29b) we obtain

$$\dot{y}_j = \sum_i^n (\partial h_j/\partial x_i)\dot{x}_i = \sum_i^n (\partial h_j/\partial x_i)\left[\sum_k^n B_{ik}(\mathbf{c}) x_k + f_i(\mathbf{x}, \mathbf{h}(\mathbf{x}); \mathbf{c})\right]$$

$$= \sum_l^m C_{jl}(\mathbf{c}) y_l + g_j(\mathbf{x}, \mathbf{h}(\mathbf{x}); \mathbf{c}) \tag{5.37}$$

where the first equality derives from the formula for differentiation of a complex function, the second derives from (5.29a) and the third equality is a consequence of (5.29b).

From (5.37) we obtain the equations enabling calculation of the function $\mathbf{h}(\mathbf{x})$:

$$\sum_i^n (\partial h_j/\partial x_i)\left[\sum_k^n B_{ik}(\mathbf{c}) x_k + f_i(\mathbf{x}, \mathbf{h}(\mathbf{x}); \mathbf{c})\right]$$

$$\sum_l^m C_{jl}(\mathbf{c}) y_l + g_j(\mathbf{x}, \mathbf{h}(\mathbf{x}); \mathbf{c}), \quad j = 1, ..., m \tag{5.38}$$

with the boundary conditions

$$h_j(0) = 0, \qquad (\partial h_j/\partial x_i)(0) = 0$$
$$j = 1, ..., m, \qquad i = 1, ..., n \tag{5.39}$$

5.4.6 Relation between the Tikhonov theorem and the centre manifold theorem

The centre manifold theorem may be regarded as a generalization of the Tikhonov theorem to a case when the time hierarchy does not explicitly occur in the examined system, i.e. when the equations cannot be divided into fast and slow. On the other hand, when the system of equations to be considered contains the time hierarchy, both the theorems are in principle equivalent (although the centre manifold theorem allows a more thorough reduction of the system). The analogy between the two systems will be exemplified by a system of equations with the marked time hierarchy (application of the centre manifold theorem to a system to which the Tikhonov theorem cannot be applied will be discussed in the next section):

$$\dot{x} = y \tag{5.40a}$$

$$\dot{y} = -x + \delta(1 - z)y \tag{5.40b}$$

$$\varepsilon\dot{z} = -z + x^2 \tag{5.40c}$$

in which $0 < \varepsilon \ll 1$ and δ is a control parameter.

The system (5.40) meets assumptions of the Tikhonov theorem. Equation (5.40c) is a fast equation while (5.40a, b) are slow equations. In accordance with the Tikhonov theorem (5.40) is replaced by the approximate system

$$\dot{x} = y \tag{5.41a}$$

$$\dot{y} = -x + \delta(1 - z)y \tag{5.41b}$$

$$0 = -z + x^2 \tag{5.41c}$$

from which we obtain the van der Pol system of equations:

$$\dot{x} = y \tag{5.42a}$$

$$\dot{y} = -x + \delta(1 - x^2)y \tag{5.42b}$$

The centre manifold method will now be applied. Linearization of (5.40) nearby its stationary point $(0, 0, 0)$, $x = 0 + \xi$, $y = 0 + \eta$, $z = 0 + \zeta$, where $|\xi|$, $|\eta|$, $|\zeta| \ll 1$, yields

$$\dot{\xi} = \eta \tag{5.43a}$$

$$\dot{\eta} = \xi + \delta\eta \tag{5.43b}$$

$$\dot{\zeta} = -(1/\varepsilon)\zeta \tag{5.43c}$$

The eigenvalues of the stability matrix are equal to (see equation (5.9))

$$\lambda_{1,2} = 1/2\delta \pm 1/2(\delta^2 - 4)^{1/2}, \quad \lambda_3 = -1/\varepsilon \tag{5.44a}$$

$$\lambda_{1,2} = 0 \pm i, \quad \lambda_3 = -1/\varepsilon < 0, \quad \text{for} \quad \delta = 0 \tag{5.44b}$$

hence, for $\delta = 0$ the system (5.40) has a sensitive state dependent on the variables x, y. Thus, for $\delta = 0$ (5.40) is of the form (5.29), where $\mathbf{x} = (x, y)$, $\mathbf{y} = z$. Accordingly, it follows from the centre manifold theorem that (5.40) may be reduced to the form

$$\dot{x} = y \tag{5.45a}$$

$$\dot{y} = -x + \delta\left[1 - h(x, y)\right] y \tag{5.45b}$$

containing complete information about sensitive states of the original system. It can be demonstrated, using equation (5.38) that, to a first approximation,

$$h(x, y) \cong x^2 \tag{5.46}$$

which leads to the system (5.42).

In Section 5.6 we will discuss a catastrophe occurring in the reduced (van der Pol) system (5.41) when the parameter δ changes sign. A catastrophe of this type: the appearance of a limit cycle, associated with a loss of stability by the stationary point $(0, 0)$, i.e. the Hopf bifurcation, also occurs in the original system (5.40). The resulting limit cycle is in this case localized on the surface $z = x^2$, see Fig. 75.

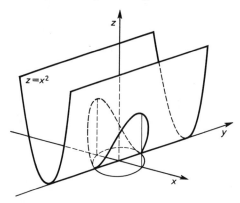

Fig. 75. Limit cycle for the system (5.40).

5.4.7 Example of application of the centre manifold theorem

We shall now show an application of the centre manifold theorem to the system without marked time hierarchy (this is the Duffing equation):

$$\dot{u} = v \tag{5.47a}$$

$$\dot{v} = \beta u - \delta v - u^2 \tag{5.47b}$$

where δ is a fixed parameter, $\delta > 0$, while β is a control parameter, $|\beta| \ll 1$.

The system (5.47) has the stationary state $(\bar{u}, \bar{v}) = (0, 0)$. A system linearized nearby the stationary point is of the form

$$\dot{\xi} = \eta \tag{5.48a}$$

$$\dot{\eta} = \beta\xi - \delta\eta \tag{5.48b}$$

where $|\xi|, |\eta| \ll 1$. The eigenvalues of the stability matrix satisfy eigenequation (5.9) and are given by

$$\lambda_{1,2} = -1/2\delta \pm 1/2[\delta^2 + 4\beta]^{1/2} \tag{5.49a}$$

$$\lambda_1 \cong \beta/\delta, \quad \lambda_2 \cong -\delta - \beta/\delta \tag{5.49b}$$

Hence, for $\delta > 0$ and $|\beta| \ll 1$ the stationary point is a saddle when $\beta > 0$ and a stable node when $\beta < 0$. The system has a sensitive state for $\beta = 0$, then $\lambda_1 = 0$, $\lambda_2 = -\delta$. When $\beta = 0$, the eigenvectors of the stability matrix, fulfilling equation (5.8), are given by the equations:

$$\lambda_1 = 0, \quad \begin{pmatrix} \xi_0 \\ \eta_0 \end{pmatrix} = \begin{pmatrix} 1 \\ 0 \end{pmatrix}, \quad \lambda_2 = 0, \quad \begin{pmatrix} \xi_0 \\ \eta_0 \end{pmatrix} = \begin{pmatrix} 1 \\ -\delta \end{pmatrix} \tag{5.50}$$

We will now perform a transformation to new variables x, y in which for $\beta = 0$ the stability matrix has a diagonal form. It is known from linear algebra that columns of the matrix transforming the variables u, v to the variables x, y having this property are constructed from the eigenvectors (5.50). Hence, the new variables x, y are of the form

$$\begin{pmatrix} u \\ v \end{pmatrix} = \begin{pmatrix} 1 & 1 \\ 0 & -\delta \end{pmatrix} \begin{pmatrix} x \\ y \end{pmatrix}, \quad \begin{pmatrix} x \\ y \end{pmatrix} = \begin{pmatrix} 1 & 1/\delta \\ 0 & -1/\delta \end{pmatrix} \begin{pmatrix} u \\ v \end{pmatrix} \tag{5.51}$$

In the new variables the Duffing equation takes the form

$$\dot{\beta} = 0 \tag{5.52a}$$

$$\dot{x} = (\beta/\delta)x + (\beta/\delta)y - (1/\delta)(x + y)^2 \tag{5.52b}$$

$$\dot{y} = -(\beta/\delta)x - (\delta + \beta/\delta)y + (1/\delta)(x + y)^2 \tag{5.52c}$$

In the new variables the main directions of the stability matrix coincide with the x, y axes. Following linearization, the system has, for $\beta = 0$, two eigenvalues equal to zero, corresponding to the β, x variables and one eigenvalue $-\delta < 0$, corresponding to the y variable. Equation (5.52a) was formally added to enable the search for the equation for the centre manifold in the form of a function in the variables x, β.

Owing to a transformation to new variables, the Duffing equation (5.52) is of the form (5.29). We will thus look for an (approximate) equation for the centre manifold in the form

$$y = h(\beta, x) = ax^2 + b\beta x + c\beta^2 + \ldots \tag{5.53}$$

where a, b, c are constants which need to be determined.

Note that a function h of this form automatically meets the requirements (5.39). The application of (5.38) leads to the equation

$$\partial h/\partial x \left[(\beta/\delta)(x + h) - (1/\delta)(x + h)^2 \right] + \partial h/\partial \beta [0]$$
$$= -\delta h - (\beta/\delta)(x + h) - (1/\delta)(x + h)^2 = 0 \tag{5.54}$$

Substituting (5.53) into (5.54) and setting to zero the coefficients of the powers β^2, βx, x^2 we obtain:

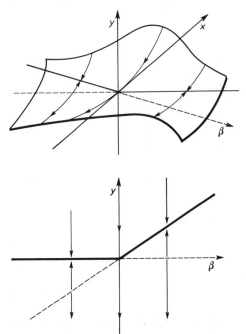

Fig. 76. Phase portraits for the systems (5.52) and (5.57).

$$a = 1/\delta^2, \quad b = -1/\delta^2, \quad c = 0 \qquad (5.55)$$

$$y = (1/\delta^2)(x^2 - \beta x) + \dots \qquad (5.56)$$

Inserting (5.56) into (5.52b) we finally obtain a reduced equation (preserving the stability properties of the overall system (5.52))

$$\dot{x} = (\beta/\delta)(1 - \beta/\delta^2)x - (1/\delta)(1 - \beta/\delta^2)x^2 + \dots \qquad (5.57a)$$

$$\dot{\beta} = 0 \qquad (5.57b)$$

where $\delta > 0$, $|\beta| \ll 1$. Equation (5.57a) corresponds to a standard form of the transcritical bifurcation, discussed in Section 5.5.2.2, which has a sensitive state of the same kind, occurring when $\beta = 0$. In the end, note that a comparison of phase diagrams for the original system (5.52) with the centre manifold marked, with those for the reduced system (5.57) reveals their qualitatively similar character (Fig. 76).

5.5 CLASSIFICATION OF CATASTROPHES IN DYNAMICAL SYSTEMS

5.5.1 Introduction

Consider a system of ordinary differential equations dependent on the parameters $\mathbf{c}$:

$$\dot{x}_i = f_i(\mathbf{x}, \mathbf{c}) \qquad (5.58)$$

where $\mathbf{x} \in R^n$ $(i = 1, \dots, n)$, $\mathbf{c} \in R^k$.

The critical (bifurcation) value of the parameters, $\mathbf{c}_0$, is defined as their values such that in the vicinity of $\mathbf{c}_0$ there is the point $\mathbf{c}_1$, such that the phase portraits determined by $\mathbf{f}(\mathbf{x}; \mathbf{c}_0)$ and $f(\mathbf{x}; \mathbf{c}_1)$ are qualitatively different. On the basis of the analysis carried out in Section 5.4.5 the critical value $\mathbf{c}_0$ is known to correspond to eigenvalues of the system of equations obtained by linearizaton of the system (5.58) for which $\mathrm{Re}(\lambda) = 0$.

Two general problems associated with a catastrophe (bifurcation) of this type can be formulated:

(1) to what extent the geometrical structure of solutions is resistant to perturbations;

(2) to what extent expansions in a series nearby a stationary (equilibrium) state may be used.

The above problems are closely related to the problem of structural stability, in analogy with the problems of elementary catastrophe theory.

We shall examine the class of bifurcation problems for which the system (5.58) has the stationary state $(\mathbf{x}_0; \mathbf{c}_0)$, that is $\mathbf{f}(\mathbf{x}_0; \mathbf{c}_0) = 0$, for which the condition $\mathrm{Re}(\lambda_1) = \ldots = \mathrm{Re}(\lambda_p) = 0$ is satisfied. Such a stationary state is sensitive. As we have established previously, a catastrophe involving a change in the phase portrait near a stationary point takes place on the system crossing a sensitive state in which the stability matrix has the eigenvalue (eigenvalues) equal to zero or a pair (pairs) of purely imaginary eigenvalues.

Recall that the condition for an occurrence of the catastrophe of a change in the phase portrait nearby a stationary point derives from the Grobman–Hartman theorem: if, on changing a control parameter, the $\mathrm{Re}(\lambda)$ value does not pass through zero, then the qualitative alteration of the phase portrait in the vicinity of a stationary point cannot occur.

Hence, the simplest cases in which a sensitive state resulting in the occurrence of a catastrophe of this type appears, are as follows:

(1) the stability matrix has one zero (real) eigenvalue or a pair of purely imaginary (conjugate) eigenvalues;

(2a) there are two zero real eigenvalues;

(2b) there is one zero real eigenvalue and a pair of purely imaginary eigenvalues;

(2c) there are two pairs of purely imaginary eigenvalues.

In cases (2) (called the cases of double degeneracy), at least two control parameters are required to describe all possible modifications of a phase portrait in the neighbourhood of a stationary state. The figure two is associated with two requirements which have to be met by the vector field $\mathbf{f}$, equation (5.61), in the case of double degeneracy; for example in case (2a) these are the conditions $P(0) = 0$, $(\mathrm{d}P/\mathrm{d}\lambda)(0) = 0$ where $P(\lambda)$ is a characteristic polynomial of the stability matrix a_{ij}. The number of independent conditions imposed on the vector field is called codimension, because in the case of the gradient field, $\mathbf{f} = \nabla_x V(\mathbf{x}; \mathbf{c})$, it is a codimension (see Chapter 2). Thus, the vector field $\mathbf{f}$ not having nearby a stationary state ($\mathbf{f} = 0$) the eigenvalues λ such that $\mathrm{Re}(\lambda) = 0$, has codimension zero (in the vector field language this is the case of a nondegenerate critical point). The sensitive states given in case (1) are of codimension one, those reported in cases (2a), (2b), (2c) are of codimension two.

A codimension of k of the vector field $f(x)$ implies that at least k control parameters c, on which the family of vector fields $f(x; c)$ must depend, are required to describe all possible modifications of the phase portrait of (5.61) nearby its stationary state. Such the family of functions $f(x; c)$ is called, by analogy with elementary catastrophe theory, a universal unfolding of codimension k, of the field $f(x)$.

Guckenheimer classified and examined the catastrophes of codimension one and two, involving a change in the phase portrait in the vicinity of a given stationary state, executing the following program (the program is a methodological and notional continuation of Thom's program for elementary catastrophe theory):

(I) the first step involves a transformation of coordinates nearby a given stationary point to the simplest form, so-called standard form;

(II) in the second step the form of a universal unfolding is found;

(III) dynamics of the system in the vicinity of the equilibrium state is determined next;

(IV) structural stability of the solutions obtained is then examined;

(V) in the case of systems with imaginary eigenvalues, the possibility of an appearance of complex resonance phenomena (which may lead, for example, to the appearance of chaotic dynamics) is studied.

The investigation of stability and catastrophes of specific dynamical systems may now proceed in the following way:

(1) the first step involves a linearization of the system and finding its sensitive states;

(2) determination of the type of a sensitive state and its codimension;

(3) application of the centre manifold theorem (an analogue of the splitting lemma in elementary catastrophe theory) to isolate the dynamics of the system which may give rise to an occurrence of catastrophes;

(4) comparison of the reduced system with known standard forms of the catastrophes of codimension equal to that of the examined system;

(5) reconstruction of dynamics of the system under study on the basis of known dynamics of the standard system.

5.5.2 Catastrophes of codimension one

We will consider dynamical systems in which the vector field f depends on one control parameter

$$\dot{x}_i = f_i(\mathbf{x}; \mathbf{c}) \tag{5.59}$$

$\mathbf{x} \in R^n$ ($i = 1, ..., n$), $c \in R^1$. Changes in stationary states and limit cycles will be examined as a function of the parameter c. A catastrophe takes place when the number of stable stationary states or limit cycles is altered.

The catastrophe of a change in the phase portrait nearby a stationary state requires (at least one) real zero eigenvalue or (at least one) purely imaginary conjugate pair of eigenvalues to be present among the eigenvalues of the stability matrix.

The centre manifold theorem generalizes the stable manifold theorem to the case of existence of eigenvalues for which $\mathrm{Re}(\lambda) = 0$. It follows from this theorem that a phase portrait is insensitive to perturbing the eigenvalues of non-zero values of $\mathrm{Re}(\lambda)$. This signifies that the case of a catastrophe in which a sensitive state is the zero real eigenvalue may be described by the vector field having one component whereas the case of a sensitive state to which corresponds a pair of imaginary eigenvalues may be qualitatively represented by equation (5.59), in which the field $\mathbf{f}$ has two components.

We will now discuss the shape of standard forms for the catastrophes of codimension one. These will be standard forms for which the function $\mathbf{f}(\mathbf{x}; \mathbf{c})$, equation (5.59), is of the form:

$$f(x; c) = c - x^2 \tag{5.60}$$

$$f(x; c) = cx - x^2 \tag{5.61}$$

$$f(x; c) = cx - x^3 \tag{5.62}$$

and the stability matrix has one zero eigenvalue and

$$\mathbf{f}(r, \varphi; c) \equiv [f_1(r, \varphi; c), \quad f_2(r, \varphi; c)] = [r(c - r^2), 1] \tag{5.63}$$

where the stability matrix has one pair of purely imaginary eigenvalues (the functions $f_{1,2}$ were expressed in polar variables: $x = r\cos(\varphi)$, $y = r\sin(\varphi)$. In all equations for standard forms given below, control parameters will be denoted as $a, b, ...$, whereas some other fixed parameters will be represented by letters $R, S, ..., P_4, Q_4$ stand for fourth-order polynomials in state variables. The sensitive state of highest degeneracy of a system is obtained by setting zero for all control parameters. The standard forms are selected in such a way that the sensitive state is always located at the origin.

5.5.2.1 Saddle node bifurcation

The standard form for a catastrophe of this type is:

$$\dot{x} = -x^2 + a \qquad (5.64)$$

Equation (5.64) is a gradient system of the potential

$$V(x; c) = 1/3 x^3 - ax \qquad (5.65)$$

The stationary state of $\bar{x}$, $\dot{\bar{x}} = 0$ or $\nabla_x V(\bar{x}; c) = 0$, given by

$$\bar{x}_{\pm} = \pm\sqrt{a}, \quad a > 0 \qquad (5.66)$$

is a degenerate critical point when

$$(\partial^2 V/\partial x^2)(\bar{x}) = 0, \quad \text{that is when} \quad a = 0 \qquad (5.67)$$

Thus, the value of the parameter $a = 0$ corresponds to a sensitive state.

Apparently, the same result is obtained by diagonalization, for $a > 0$, of equation (5.64) near the stationary point (5.67) and computation of the (one-dimensional) eigenvalue of the stability matrix:

$$x = \bar{x} + \xi, \quad |\xi| \ll 1 \qquad (5.68a)$$

$$\dot{\xi} = -\left(\pm 2\sqrt{a}\right)\xi \qquad (5.68b)$$

$$\xi = \exp(\lambda t) \qquad (5.68c)$$

$$\lambda_{\pm} = \mp\left(2\sqrt{a}\right) \qquad (5.68d)$$

The sensitive state, $\lambda = 0$, corresponds to the value of the parameter $\lambda = 0$. Moreover, we can see that the stationary state $\bar{x}_+$ is stable (state of rest) and the stationary state $\bar{x}_-$ is unstable.

When the parameter a approaches zero from the side of positive numbers, a catastrophe takes place: both the stationary points vanish. The phase portraits for equation (5.64) for $a > 0$, $a = 0$, $a < 0$, are shown in Fig. 77.

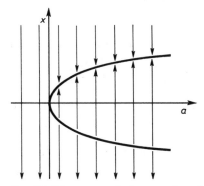

Fig. 77. Phase portrait for equation (5.64) for $a > 0$, $a = 0$, $a < 0$.

Fig. 78. Bifurcation diagram for equation (5.64).

The following figure (Fig. 78) depicts a so-called bifurcation diagram, that is the plot of stationary states $\bar{x}(c)$ vs. the parameter c. A dashed line represents an unstable state.

The problem remains to be decided when a catastrophe of this type may appear in a general system of the type (5.59). The problem has been resolved by the following theorem.

Let $\bar{x}(c)$ be the stationary state of the system (5.59) for which the following conditions are fulfilled:

(1) the stability matrix has the single eigenvalue $\lambda = 0$ for $c = c_0$, while the remaining eigenvalues lie outside the imaginary axis;
(2) the system does not have the stationary state $\bar{x} = 0$;
(3) the symmetry conditions are not imposed on the system.

Upon satisfying the above conditions, a saddle node bifurcation occurs for $c = c_0$.

5.5.2.2 Transcritical bifurcation

When condition (2) in the above theorem is not met, a transcritical bifurcation may take place.

The standard form for a catastrophe of this type is:

$$\dot{x} = -x^2 + ax \tag{5.69}$$

We shall examine properties of the system (5.69) using only the approach based on investigation of eigenvalues of the stability matrix.

Stationary states of the system (5.69), $\bar{x} = 0$, are of the form

$$\bar{x}_1 = 0 \tag{5.70a}$$

$$\bar{x}_2 = a \tag{5.70b}$$

Linearization of the equation in the neighbourhood of $\bar{x}_1$, $\bar{x}_2$ yields

$$x \cong \bar{x}_1 + \xi_{1,0}\exp(at) \tag{5.71}$$

$$x \cong \bar{x}_2 + \xi_{2,0}\exp(-at) \tag{5.72}$$

The stationary state $\bar{x}_1 = 0$ is trivial inasmuch as it occurs for each value of the control parameter. A sensitive state appears when $a = 0$; then $\lambda_1 = = \lambda_2 = 0$.

Let now the control parameter a vary continuously, passing from negative to positive values. When $a < 0$, then the stationary state $\bar{x}_1 = 0$ is stable and the stationary state $\bar{x}_2 = a$ is unstable; for $a = 0$ the stationary state $\bar{x}_1 = \bar{x}_2 = 0$ is a sensitive state; when $a > 0$ the stationary state $\bar{x}_1$ is unstable while the state $\bar{x}_2 = a$ is stable. A phenomenon of this type is called the stability exchange.

The phase portraits for equations (5.69) for $a > 0$, $a = 0$, $a < 0$ are shown in Fig. 79.

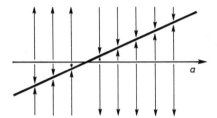

Fig. 79. Phase portrait for equation (5.69) for $a > 0$, $a = 0$, $a < 0$.

Fig. 80. Bifurcation diagram for equation (5.69).

The next figure (Fig. 80) illustrates a bifurcation diagram, that is the plot of the stationary states $\bar{x}(c)$ as a function of the parameter c. The unstable state is represented by a dashed line.

5.5.2.3 Pitchfork bifurcation

In this case the system is symmetric under reflection:

$$f(x) = +f(-x) \tag{5.73a}$$

$$f(x) = -f(-x) \tag{5.73b}$$

The simplest system satisfying condition (5.73b) is the following standard form:

$$\dot{x} = -x^3 + ax \tag{5.74}$$

The stationary states of the system (5.74), $\dot{x} = 0$, are of the form

$$\bar{x}_0 = 0 \tag{5.75}$$

$$\bar{x}_{\pm} = \pm\sqrt{a}, \quad a > 0 \tag{5.76}$$

Linearization of the equation in the neighbourhood of $\bar{x}_0$, $\bar{x}_{\pm}$ yields successively

$$x = \bar{x}_0 + \exp(at) \tag{5.77a}$$

$$x = \bar{x}_+ + \exp(-2at) \tag{5.77b}$$

$$\lambda_+ = -2a \tag{5.78a}$$

$$x = \bar{x}_- + \exp(2at) \tag{5.78b}$$

The stationary state appears when $a = 0$; then $\lambda_0 = \lambda_+ = \lambda_- = 0$.

Let now the control parameter a vary continuously, passing from positive to negative values. When $a > 0$, then the stationary state $\bar{x}_0 = 0$ is stable, the stationary state $\bar{x}_+ = +\sqrt{a}$ is stable and the state $\bar{x}_- = -\sqrt{a}$ is unstable; on the other hand, for $a < 0$ the only stable state is the stationary state $\bar{x}_0$.

The phase portraits for equation (5.74) for $a > 0$, $a = 0$, $a < 0$ are shown in Fig. 81. In turn, Fig. 82 illustrates a bifurcation diagram, i.e. the

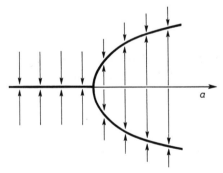

Fig. 81. Phase portrait for equation (5.74) for $a > 0$, $a = 0$, $a < 0$.

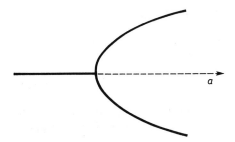

Fig. 82. Bifurcation diagram for equation (5.74).

dependence of stationary states $x(c)$ on the parameter c. A dashed line denotes an unstable state.

5.5.2.4 Hopf bifurcation

According to equation (5.63d), the standard form of the system has the following form (in polar variables r, φ, $r \geqslant 0$);

$$\dot{r} = r(c - r^2) \tag{5.79a}$$

$$\dot{\varphi} = 1 \tag{5.79b}$$

On converting to Cartesian coordinates, $x = r\cos(\varphi)$, $y = r\sin(\varphi)$ we obtain an alternative (although more complex) standard form

$$\dot{x} = cx - y - x(x^2 + y^2) \tag{5.80a}$$

$$\dot{y} = x + cy - y(x^2 + y^2) \tag{5.80b}$$

The system (5.80) has only one stationary state $\dot{x} = \dot{y} = 0$ given by

$$(\bar{x}, \bar{y}) = (0, 0) \tag{5.81}$$

that is $\bar{r} = 0$.

Linearization of equations (5.80) nearby the stationary state yields:

$$x \cong \xi = \xi_0 \exp(\lambda t) \tag{5.82a}$$

$$y \cong \eta = \eta_0 \exp(\lambda t) \tag{5.82b}$$

where ξ, η satisfy the linearized equation

$$\dot{\xi} = c\xi - \eta \tag{5.83a}$$

$$\dot{\eta} = \xi + c\eta \tag{5.83b}$$

Hence, the eigenvalues fulfil the characteristic equation

$$\det \begin{vmatrix} (c - \lambda) & -1 \\ 1 & (c - \lambda) \end{vmatrix} = (c - \lambda)^2 + 1 = 0 \tag{5.84a}$$

$$\lambda_{1,2} = c \pm i \tag{5.84b}$$

When the parameter $c < 0$ the stationary state $(\bar{x},\bar{y}) = (0, 0)$ is of a stable focus type; for $c = 0$ we deal with the sensitive state: the stationary state is a centre; for $c > 0$ a catastrophe takes place, because the stationary state becomes an unstable focus. The phase trajectories for the linearized system (5.83) for $c < 0$, $c = 0$, $c > 0$ nearby the stationary point are shown in Figs. 65, 66, 67, respectively.

It remains to be settled what happens to the trajectories escaping from the unstable neighbourhood (for $c > 0$) of a stationary point. The question may be readily answered by examining the standard form (5.79). We can see that for $c > 0$ there exists, besides the stationary state $\bar{r} = 0$, also a limit cycle of the form

$$\bar{r} = +\sqrt{c}, \quad \varphi = t \tag{5.85}$$

$$x(t) = \bar{r}\cos(t), \quad y(t) = \bar{r}\sin(t) \tag{5.86}$$

When $r(t) > \bar{r}$ the right-hand side of equation (5.79a) is negative and $r(t)$ decreases; when $r(t) < \bar{r}$ the right-hand side of (5.79a) is positive and $r(t)$ increases. Consequently, the limit cycle (5.85), (5.86) is stable (attracting). We may thus plot the phase portrait of system (5.79) or (5.80) in the entire phase space, see Fig. 68a.

A catastrophe of this type is called the Hopf bifurcation. It may be represented by a bifurcation diagram in which the position of a stationary state and a limit cycle are plotted as a function of the parameter c (Figs. 83, 84).

In a general case of equation (5.62), the necessary condition on the Hopf bifurcation is that the stability matrix has, for the critical value of the control parameter $c = c_0$, one pair of purely imaginary eigenvalues, whereas the remaining eigenvalues must have a non-zero real part.

The Hopf bifurcation is a dynamical catastrophe, since a stable stationary state bifurcates to a limit cycle; hence, state variables change with time. Interestingly, the Hopf bifurcation may be visualized by elementary catastrophe theory despite the fact that the Hopf bifurcation may not appear in a gradient system and, furthermore, it is a dynamical catastrophe.

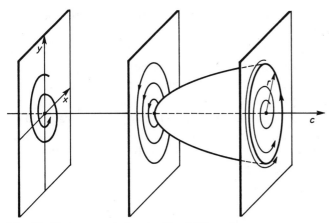

Fig. 83. Phase portrait for the Hopf bifurcation.

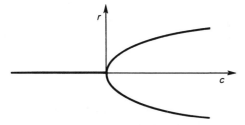

Fig. 84. Bifurcation diagram for the Hopf bifurcation catastrophe.

Such an analysis of the Hopf bifurcation (and some other catastrophes) will be presented in Section 5.6.

There are also catastrophes of codimension one not leading to a qualitative change of the stationary state. These are catastrophes of a global type. An example of the sensitive state corresponding to a global catastrophe is the state $\lambda_1 = \lambda_2$. The phase trajectories for such sensitive states in linear systems were given in Section 5.2.

5.5.2.5 Structural stability of catastrophes of codimension one

The saddle node catastrophe and the Hopf bifurcation may be shown to be structurally stable. Certain additional conditions (see Sections 5.5.2.2, 5.5.2.3) are imposed on the transcritical bifurcation and the pitchfork bifurcation. The system is structurally stable under perturbations not disturbing these additional conditions; on the other hand, when arbitrary perturbations are allowed, the catastrophes are structurally unstable. The implies two possibilities.

The examined physical system always satisfies stipulated additional conditions (an additional stationary point or suitable symmetry) and then the described catastrophe may appear in such systems in a structurally stable way. If the examined system does not fulfil these conditions, the description of the system is inadequate and a possibility of the description of a catastrophe of higher codimension should be considered. For example, such a structurally stable extension of the pitchfork bifurcation is the cusp catastrophe described in Section 5.5.3.2.

5.5.3 Catastrophes of codimension two

5.5.3.1 Introduction

We shall consider dynamical systems in which the vector field $\mathbf{f}$ depends on two control parameters

$$\dot{x}_i = f_i(\mathbf{x}; \mathbf{c}) \tag{5.87}$$

$\mathbf{x} \in R^n \ (i = 1, ..., n), \ c \in R^2$.

It can be shown that there exist five nonequivalent sensitive states of codimension two. Their standard forms will be discussed below. The catastrophes have common names; we shall also introduce nomenclature (reflecting the type of a sensitive point).

The stability matrices corresponding to these standard forms have successively: (1) one zero eigenvalue; (2) one pair of purely imaginary eigenvalues; (3) two eigenvalues equal to zero; (4) one zero eigenvalue and one pair of purely imaginary eigenvalues; (5) two different pairs of purely imaginary eigenvalues.

5.5.3.2 Cusp catastrophe (0)

The following standard form corresponds to this catastrophe:

$$\dot{x} = x^3 + bx + a \tag{5.88}$$

This is a generalization of the catastrophe of a saddle node type or, more specifically, of a pitchfork catastrophe. The sensitive state corresponds to one zero eigenvalue. The catastrophe is easy to describe, being of a gradient type. It appears when the term x^3 occurs on the right-hand side of equation (5.88) and the requirement on the pitchfork bifurcation of an occurrence of the additional stationary state $\bar{x} = 0$ is not satisfied (i.e. the parameter a cannot be ignored).

5.5.3.3 Degenerate Hopf bifurcation ($\pm i\beta$)

The standard form is the following (in polar coordinates r, φ, $r \geqslant 0$);

$$\dot{r} = r\left(c_1 + c_2 r^2 + r^5\right) \tag{5.79a'}$$

$$\dot{\varphi} = 1 \tag{5.79b'}$$

The degenerate Hopf bifurcation takes place when the term r^3 is missing from the standard form of the Hopf bifurcation (5.79). The sensitive state is the same as that for the Hopf bifurcation: one pair of purely imaginary eigenvalues $\pm i\beta$.

5.5.3.4 Takens–Bogdanov bifurcation (0, 0)

The following standard form corresponds to this catastrophe:

$$\dot{x} = y \tag{5.89a}$$

$$\dot{y} = c_1 + c_2 + R_1 x^2 + R_2 xy \tag{5.89b}$$

The sensitive state, obtained for $c_1 = c_2 = 0$, corresponds to the case $\lambda_1 = \lambda_2 = 0$, the stability matrix being not proportional to the identity matrix. It is thus case (i2) described in Section 5.2; the respective phase trajectories are shown in Fig. 70. The standard form (5.89) takes into account all possibilities of modification of phase trajectories of this sensitive state after accounting for a perturbation.

Stationary states are determined from the equations $\dot{\bar{x}} = \dot{\bar{y}} = 0$, obtaining (to simplify the equations, $R_1 = R_2 = 1$ were substituted in (5.89b))

$$\bar{x}^2 + c_2 \bar{x} + c_1 = 0, \quad \bar{y} = 0 \tag{5.90a}$$

$$\bar{x}_{1,2} = -1/2 c_2 \pm \left(c_2{}^2 - 4c_1\right)^{1/2} \tag{5.90b}$$

Linearization in the vicinity of the stationary state, $x = \bar{x} + \xi$, $y = \eta$, $|\xi|$, $|\eta| \ll 1$, leads to the linear system

$$\dot{\xi} = \eta \tag{5.91a}$$

$$\dot{\eta} = \left(c_2 + 2\bar{x}\right)\xi + \bar{x}\eta \tag{5.91b}$$

whose eigenequation (5.9) is of the form

$$\lambda^2 - \bar{x}\lambda - (c_2 + 2\bar{x}) = 0 \tag{5.91c}$$

$$\lambda_{1,2} = 1/2\bar{x} \pm [\bar{x}^2 - 4(c_2 + 2\bar{x})]^{1/2} \tag{5.91d}$$

The requirement for an occurrence of the saddle bifurcation (the sensitive state: $\lambda_1 = 0$, λ_2 different from zero) is obtained, in view of equations (5.90a) and (5.91c), in the form of a condition for the control parameters c_1, c_2 (for example, from (5.91c) follows $2\bar{x} + c_2 = 0$, $\bar{x}$ different from zero):

$$c_1 - (c_2/2)^2 = 0 \tag{5.92a}$$

The condition for an appearance of the Hopf bifurcation (the sensitive state $\lambda_{1,2} = \pm i\beta$) is obtained, on the basis of equations (5.90a), (5.91c), as the condition for control parameters (from (5.91c) follows $\bar{x} = 0$, $(2\bar{x} + c_2) > 0$):

$$c_1 = 0, \quad c_2 < 0 \tag{5.92b}$$

In the bifurcation diagram shown in Fig. 85, the plane of control parameters was divided into regions of a qualitatively different character of phase trajectories (the shapes of these trajectories are given in the respective regions) and the lines on which occur sensitive states of the Hopf bifurcation and the saddle bifurcation were marked. The diagram also depicts the line of sensitive states of the global bifurcation: the appearance of a cycle from the branches of saddle separatrices.

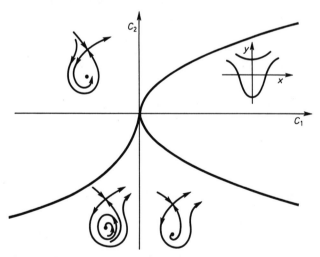

Fig. 85. Phase portraits and a bifurcation diagram for the Takens–Bogdanov bifurcation.

5.5.3.5 Takens–Bogdanov bifurcation (0, 0) *with the additional stationary state* $\bar{x} = 0$

The catastrophe has the following standard form:

$$\dot{x} = y \tag{5.89a'}$$

$$\dot{y} = c_1 x + c_2 y + R_1 x^2 + R_2 xy \tag{5.89b'}$$

A description of dynamics of this as well as the next systems can be found in a paper of Guckenheimer or in a book by Arnol'd.

5.5.3.6 Takens–Bogdanov bifurcation (0, 0) *with symmetry under rotation through the angle* π

The following standard form corresponds to this catastrophe:

$$\dot{x} = y \tag{5.89a''}$$

$$\dot{y} = c_1 x + c_2 y + R_1 x^3 + R_2 x^2 y \tag{5.89b''}$$

5.5.3.7 Pitchfork–Hopf bifurcation (0, $\pm$i)

The catastrophe has the corresponding standard form:

$$\dot{\varphi} = a + R_1 r^2 \tag{5.93a}$$

$$\dot{r} = b_1 r + R_2 r x_3 + \left(S_1 r^3 + S_2 x_3^2\right) \tag{5.93b}$$

$$\dot{x}_3 = b_2 + R_3 x_3^2 + R_4 r^2 + S_3 r^2 x_3 + S_4 x_3^3 \tag{5.93c}$$

5.5.3.8 Pitchfork–Hopf bifurcation (0, $\pm$i) *with symmetry under reflection* $x_3 \to -x_3$

The catastrophe has the corresponding standard form:

$$\dot{\varphi} = a + R_1 r^2 \tag{5.94a}$$

$$\dot{r} = b_1 r + S_1 r x_3^2 + S_2 r^3 + r P_4 \tag{5.94b}$$

$$\dot{x}_3 = b_2 x_3 + S_3 r^2 x_3 + S_4 x_3^3 + x_3 Q_4 \tag{5.94c}$$

5.5.3.9 Double Hopf bifurcation ($\pm$iα, $\pm$iβ) *without resonance*

The following standard form corresponds to this catastrophe:

$$\dot{\varphi}_1 = a_1 + R_1 r_1^2 + R_2 r_2^2 \tag{5.95a}$$

$$\dot{\varphi}_2 = a_2 + R_3 r_1^2 + R_2 r_2^2 \tag{5.95b}$$

$$\dot{r}_1 = r_1 \left(b_1 + S_1 r_1^2 + S_2 r_2^2 + P_4\right) \tag{5.95c}$$

$$\dot{r}_2 = r_2\left(b_2 + S_3 r_1{}^2 + S_4 r_2{}^2 + Q_4\right) \tag{5.95d}$$

5.5.4 Catastrophes of codimension three

A complete classification of the catastrophes in two state variables of codimension three has not yet been developed. Consequently, only the standard form of a catastrophe corresponding to the sensitive state of the Takens–Bogdanov bifurcation will be given. The generalized Takens––Bogdanov bifurcation has the corresponding standard form:

$$\dot{x} = y \tag{5.89a$'$}$$

$$\dot{y} = c_1 + c_2 x + c_3 x^2 \pm x^3 + R_1 xy + R_2 y^2 + R_3 x^2 y \tag{5.89b$'$}$$

A description of dynamics and bifurcation diagrams for this catastrophe have been given by Medved.

5.6 ELEMENTARY ANALYSIS OF DYNAMICAL CATASTROPHES

We will now describe the Arnol'd method of examination of the stability of dynamical systems which may be regarded as a direct generalization of elementary catastrophe theory. The method is particularly well suited for analyzing the dynamical systems which may be considered to be perturbed systems having the Lyapunov function. The procedure may be employed to investigate the stability of a phase portrait nearby a stationary point more exactly than by using the linear approximation described in Section 5.1; the results obtained pertain to a considerably larger neighbourhood of the stationary point and are suitable for the examination of dynamical catastrophes.

A description of the method will be begun with the analysis of a catastrophe occurring in the van der Pol system

$$\dot{x} = y \tag{5.96a}$$

$$\dot{y} = -x + \varepsilon y\left(1 - x^2\right) \tag{5.96b}$$

where the parameter ε is small, $|\varepsilon| < 1$. The system (5.96) has the stationary point (0, 0). Linearization of (5.96) in the vicinity of this point leads to the system of equations

$$\dot{x} = y \tag{5.97a}$$

$$\dot{y} = -x + \varepsilon y \tag{5.97b}$$

The eigenvalues of this system (see (5.9)) are given by

$$\lambda_{1,2}(\varepsilon) = 1/2\varepsilon \pm 1/2(\varepsilon^2 - 4)^{1/2} \tag{5.98}$$

Hence, the stationary point $(0, 0)$ for small (as regards the absolute value) ε is a stable focus and for small positive ε is an unstable focus. When the parameter ε changes sign, a catastrophe — a change in the nature of trajectories, takes place in the system. In addition, $\lambda_{1,2}(0) = \pm i$; hence, the state of the system corresponding to $\varepsilon = 0$ is a sensitive state typical for the Hopf bifurcation.

The system (5.96) may be regarded as a perturbed Hamiltonian system

$$\dot{x} = y \tag{5.99a}$$

$$\dot{y} = -x \tag{5.99b}$$

whose Hamiltonian is equal to

$$H = 1/2(x^2 + y^2) \tag{5.100}$$

The Hamiltonian function is constant on trajectories of the system (5.99)

$$\overset{\circ}{H}(x, y) = x(y) + y(-x) = 0 \tag{5.101}$$

where we employed equation (A16), i.e. the phase trajectories are circles given by the equation $H = $ const. Thus, the stationary point $(0, 0)$ of the system (5.99) is of a centre type.

Since ε is a small parameter, it may be presumed that, to a first approximation, the function H also corresponds to the energy of the perturbed system (5.96). Let us thus compute how the function H varies with time on the trajectories of the perturbed system (5.96). In view of equations (A.16), (5.96) we obtain

$$\overset{\circ}{H}(x, y) = x(y) + y[-x + \varepsilon y(1 - x^2)] = \varepsilon y^2(1 - x^2) \tag{5.102}$$

We may now calculate the change in energy of the system, ΔH, in the course of its evolution along the phase trajectory:

$$\Delta H = \int_{t_1}^{t_2} \overset{\circ}{H}[x(t), y(t)] \, dt \tag{5.103}$$

Not knowing the exact trajectories of the system (5.96) we may determine an approximate value of ΔH by computing the integral (5.103) on the trajectories of the system (5.99), i.e. on the circles of radius R

$$x = R\cos(t) \tag{5.104a}$$

$$y = -R\sin(t) \tag{5.104b}$$

$$H = 1/2\,R^2 \tag{5.104c}$$

Hence, the ΔH value calculated for one revolution is equal to

$$\Delta H = \int_0^{2\pi} \left[\varepsilon y^2(1 - x^2)\right] dt = \varepsilon \int_0^{2\pi} R^2\sin^2(t)\left[1 - R^2\cos^2(t)\right] dt$$

$$= \varepsilon\pi R^2 \left(1 - 1/4R^2\right) \tag{5.105}$$

Thus, the change of energy of the system (5.96) per revolution, given by an approximate equation (5.105), can be positive or negative. For small R values (the trajectory then occurs near the stationary point $(0, 0)$) ΔH is positive.

It follows from (5.104c) that a rise in the energy is accompanied by an increase in the distance of the phase trajectory from the unstable (for $\varepsilon > 0$) stationary point $(0, 0)$. As the type of the point is an unstable focus, the trajectory corresponds to an unwinding spiral. When $R = 2$, the change in energy per revolution is zero — the point in the phase space, representing the state of the system and moving in a circle of radius $R = 2$, has constant energy, like a Hamiltonian system. When $R > 2$, then $\Delta H < 0$ and the system loses energy; the trajectories of the system tend in a spiral to the circle of radius $R = 2$. Since our considerations were approximate, the exact limit cycle being generated in the system (5.96) does not coincide with the circle $R = 2$, see Fig. 86.

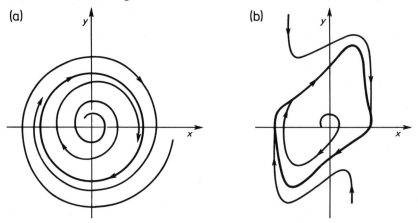

Fig. 86. Limit cycles for the van der Pol system: (a) $\varepsilon = 0.1$; (b) $\varepsilon = 1.0$.

Evidently, for $\varepsilon > 0$ a stable limit cycle appears in the system. A catastrophe of this type is called the Hopf bifurcation. Interestingly, the system can be assigned a certain energy or, more specifically, the change in energy per revolution, ΔH. The states of the system approach the closed trajectory for which $\Delta H = 0$. Hence, we may introduce a potential function V, dependent on the state parameter x, $x \equiv R^2$, such that the state which is approached by all trajectories corresponds to the condition of minimum of this potential

$$V(x) = \text{const} + 1/2\, x^2 - 1/12 x^3, \quad x \equiv R^2 \tag{5.106a}$$

$$dV/dx = x(1 - 1/4\,x) \tag{5.106b}$$

The equation $dV/dx = 0$ yields as solutions $R = 0$, i.e. the (unstable) stationary point $(0, 0)$, and the stable limit cycle, $R = 2$ (compare equations (5.106b), (5.105)).

The presented Arnol'd method may be generalized in several directions. First of all, consider the dynamical system (5.2a). The stationary points $(\bar{x}, \bar{y})$ satisfy equation (5.4). To simplify notation let us assume that the stationary point of the system (5.2) is the origin, that is the point $(0, 0)$. The system (5.2) may then be written in the form

$$\dot{x} = a_{11}x + a_{12}y + f(x, y) \tag{5.107a}$$

$$\dot{y} = a_{21}x + a_{22}y + g(x, y) \tag{5.107b}$$

where the functions $f(x, y)$, $g(x, y)$ do not contain a linear part (their expansions in a Taylor series in the vicinity of the point $(0, 0)$ begin with the terms x^2, xy, y^2).

Nearby the stationary point the functions f, g are small and the system (5.107a) may thus be regarded as a perturbed linear system

$$\dot{x} = a_{11}x + a_{12}y \tag{5.108a}$$

$$\dot{y} = a_{21}x + a_{22}y \tag{5.108b}$$

In Section 5.1 we rejected the functions f, g and studied the stability of a stationary point by examining the system (5.6) (to which (5.108) is equivalent for $\bar{x} = 0$, $\bar{y} = 0$). We will now perform a more thorough analysis of the stability of a stationary point not neglecting the functions f, g but only assuming that they are small (this implies examination of the system (5.107) in a certain vicinity of the stationary point). To emphasize this assumption we shall introduce a small parameter ε, writing the system (5.107) in the following form:

$$\dot{x} = a_{11}x + a_{12}y + \varepsilon f(x, y) \tag{5.109a}$$

$$\dot{y} = a_{21}x + a_{22}y + \varepsilon g(x, y) \tag{5.109b}$$

where $|\varepsilon| \leqslant 1$. Apparently, there are systems, such as e.g. the van der Pol system (5.96), in which a small parameter ε is preset.

Assume that we want to study the time evolution of a certain function $F(x, y)$, where x, y satisfy equations (5.109). The function F may be, for example, the Lyapunov function W, see Appendix A5. In a special case F can be a potential function or a Hamiltonian function as in the example of the van der Pol system (5.96).

We thus compute the derivative of the function F with respect to time on the trajectories of the system (5.109)

$$\overset{\circ}{F}(x, y) = \partial F/\partial x [a_{11}x + a_{12}y + \varepsilon f(x, y)] +$$
$$+ \partial F/\partial y [a_{21}x + a_{22}y + \varepsilon g(x, y)] \tag{5.110}$$

An approximate change in the quantity ΔF during time evolution may now be determined by computing the integral

$$\Delta F = \int_{t_1}^{t_2} \overset{\circ}{F}[x(t), y(t)] \, dt \tag{5.111}$$

where $[x(t), y(t)]$ is the trajectory (solution) of the unperturbed system (5.108). For small ε, thus computed quantity ΔF deviates only slightly from the exact value. As the unperturbed system (5.108) is linear, finding its solutions and the shape of trajectories present no difficulty.

In Chapter 6 we will apply the method described above to the examination of stability of some chemical kinetics equations. Moreover, in the case of establishing the existence of a sensitive state, characteristic of the Hopf bifurcation, the presence of a limit cycle may sometimes be proved (without giving its more detailed characteristic) in a different way. For this purpose suffice it to demonstrate that the trajectories of a system cannot escape to infinity and remain in some limited region. In such a case, a limit cycle must exist inside this region.

To end with, we shall discuss other possibilities of a generalization of the above results. Instead of equations (5.108), (5.109) one may study the systems

$$\dot{x} = p(x, y) \tag{5.112a}$$

$$\dot{y} = q(x, y) \tag{5.112b}$$

$$\dot{x} = p(x, y) + \varepsilon f(x, y) \tag{5.113a}$$

$$\dot{y} = q(x, y) + \varepsilon g(x, y) \tag{5.113b}$$

if the Lyapunov function and the phase trajectories of the unperturbed system (5.112) are known. Secondly, the above results may be generalized to the case of a larger of equations.

5.7 METHODS OF EXAMINATION OF EQUATIONS OF REACTIONS WITH DIFFUSION

5.7.1 Introduction

We shall investigate equations of reactions with diffusion of the form

$$\partial x_i / \partial t = f_i [\mathbf{x}(t, r); \mathbf{c}] + D_i \partial^2 x_i / \partial r^2 \tag{5.114}$$

where r is the coordinate along the longitudinal axis of a reactor, D_i are constant diffusion coefficients (independent of $\mathbf{x}$), $\mathbf{x} \in R^n$, $\mathbf{c} \in R^k$.

Note that in a general case the effects associated with diffusion in three spatial dimensions $\mathbf{r}$, the dependence of diffusion coefficients on the $\mathbf{r}$, $\mathbf{x}$ variables and the presence of (non-zero) off-diagonal diffusion coefficients should be accounted for in equations of reactions with diffusion.

In this case equations (5.114) become

$$\partial x_i / \partial t = f_i [\mathbf{x}(t, \mathbf{r}); \mathbf{c}] + \sum_j \nabla_{\mathbf{r}} (D_{ij} \nabla_{\mathbf{r}} x_j) \tag{5.114'}$$

The following solutions to the system of equations (5.114) can be distinguished: (1) states independent of r and independent of t (spatially homogeneous stationary states); (2) states independent of r, dependent on t (spatially homogeneous states); (3) states dependent on r, independent of t (stationary states); (4) states dependent on r, dependent on t.

The first step in the examination of the system (5.114) involves finding spatially homogeneous stationary states $\bar{x}_i = \text{const}$, satisfying the equations

$$f_i(\mathbf{x}; \mathbf{c}) = 0 \tag{5.115}$$

Further analysis will be exemplified by the system of two equations:

$$\partial x/\partial t = P(x, y; \mathbf{c}) + D_x(\partial^2 x/\partial r^2) \tag{5.116a}$$

$$\partial y/\partial t = Q(x, y; \mathbf{c}) + D_y(\partial^2 y/\partial r^2) \tag{5.116b}$$

5.7.2 Stability of spatially homogeneous stationary solutions to a system of equations with diffusion

The spatially homogeneous stationary states of equations (5.116) fulfil the relations

$$P(\bar{x}, \bar{y}) = 0, \quad Q(\bar{x}, \bar{y}) = 0 \tag{5.117}$$

Stability of the solutions $\bar{x}$, $\bar{y}$ will be studied. We will impose on the solutions $x(t, r)$, $y(t, r)$ of the system (5.114) the following boundary conditions for $t \geqslant 0$:

$$x(t, 0) = x(t, R) = \bar{x}, \quad y(t, 0) = y(t, R) = \bar{y} \tag{5.118}$$

where we assumed that the points on the r-axis having coordinates $r = 0$, $r = R$ correspond to the ends of the reactor (boundary conditions (5.118) correspond to an assumption that the reaction is in the stationary state at the ends of the reactor).

It should be decided what kinds of dynamics may be generated in the system when $t \to \infty$, depending on the values of parameters $\mathbf{c}$, D_x, D_y when the solutions $\bar{x}$, $\bar{y}$ lose stability.

The solutions to the system of equations (5.116) will be represented in the form

$$x(t, r) = \bar{x} + \xi(t, r) \tag{5.119a}$$

$$y(t, r) = \bar{y} + \eta(t, r) \tag{5.119b}$$

Insertion of (5.119) into (5.116) enables, with the assumption $|\xi|, |\eta| \ll 1$, its linearization

$$\partial \xi/\partial t = a_{11}\xi + a_{12}\eta + D_x(\partial^2 \xi/\partial r^2) \tag{5.120a}$$

$$\partial \eta/\partial t = a_{21}\xi + a_{22}\eta + D_y(\partial^2 \eta/\partial r^2) \tag{5.120b}$$

where a_{ij} is the stability matrix of the diffusionless system.

Deviations from the state $(\bar{x}, \bar{y})$ can be represented in terms of superposition of waves satisfying conditions (5.118):

$$\xi(t,r) = \sum_n \xi_n(t) \sin(n\pi r/R) \tag{5.121a}$$

$$\eta(t,r) = \sum_n \eta_n(t) \sin(n\pi r/R) \tag{5.121b}$$

where $n\pi/R \equiv k_n$ is the wave vector, associated with the wavelength l_n through the relationship $l_n = 2\pi/k_n$.

Substitution of (5.121) in (5.120) yields the eigensystem

$$\Lambda \xi_n = (a_{11} - D_x k_n^2)\xi_n + a_{12}\eta_n \tag{5.122a}$$

$$\Lambda \eta_n = a_{21}\xi_n + (a_{22} - D_y k_n^2)\eta_n \tag{5.122b}$$

$k_n \equiv (n\pi/R)$, analogous with the system (5.8).

Hence, the stability of a spatially homogeneous stationary solution to equations of a reaction with diffusion $(\bar{x}, \bar{y})$, may be examined in exactly the same way as we have investigated the stability of the solutions $(\bar{x}, \bar{y})$ to the system (5.2).

The vanishing of the determinant of the system of equations (5.122) is the condition for an existence of the non-zero solutions ξ_n, η_n, and this leads to the characteristic equation

$$\Lambda^2 - [a_{11} + a_{22} - (D_x + D_y)k_n^2]\,\Lambda +$$
$$+ [(a_{11} - D_x k_n^2)(a_{22} - D_y k_n^2) - a_{12}a_{21}] = 0 \tag{5.123}$$

Spatially homogeneous stationary states $(\bar{x}, \bar{y})$ may be classified in the same way as in the case of stationary states of the equations not accounting for diffusion, see Section 5.1.

The sensitive state of the system corresponds to the eigenvalues Λ meeting the requirement $\mathrm{Re}(\lambda_{1,2}) = 0$. We shall investigate the appearance of sensitive states depending on the diffusion coefficients, D_x, D_y, and on the control parameters on which the elements of the stability matrix a_{ij} rely.

The quadratic equation (5.123) has the solutions given by

$$\Lambda_{1,2}(k_n^2) = -A/2 \pm 1/2[A^2 - 4B]^{1/2} \tag{5.124a}$$

$$A = [a_{11} + a_{22} - (D_x + D_y)k_n^2] \tag{5.124b}$$

$$B = [(a_{11} - D_x k_n^2)(a_{22} - D_y k_n^2) - a_{12}a_{21}] \tag{5.124c}$$

The condition for an occurrence of the sensitive state $\mathrm{Re}(\Lambda) = 0$ can be accomplished in two ways:

$$A = 0, \quad A^2 - 4B < 0, \quad \text{that is} \quad \Lambda_{1,2} = \pm i\beta \tag{5.125a}$$

$$B = 0 \qquad\qquad \text{that is} \quad \varLambda_1 = 0 \qquad\qquad (5.125\text{b})$$

When the spatially homogeneous stationary state $(\bar{x}, \bar{y})$ loses stability and crosses a sensitive state, then in case (5.125a) travelling waves of a finite amplitude may be generated while in case (5.125b) stationary periodical spatial structures (dissipative structures) may be generated.

Several types of the dependence of $\varLambda_{1,2}$ on the parameter k^2 and the respective sensitive states may be shown to exist. Note that for very large k^2 equation (5.123) has the solutions

$$\varLambda_1 \cong -D_x k^2, \quad \varLambda_2 = -D_y k^2, \quad k^2 \gg 1 \qquad\qquad (5.126)$$

i.e. there always exists such a value of k that the solutions with eigenvalues having a negative real part and stable for $t \to \infty$ are possible.

The plots of the dependence of $\varLambda_{1,2}$ on k^2, drawn on the basis of equations (5.124), will be presented for two typical cases.

The plot $\varLambda_{1,2}(k^2)$ for the first case is illustrated in Fig. 87. The eigenvalues are initially conjugate but for a certain large k^2 they become real (we have shown above that for a sufficiently large k^2 the eigenvalues become real).

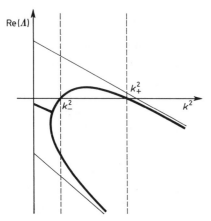

Fig. 87. Dependence of $\varLambda$ on k^2, first case.

The system has the spatially homogeneous stationary point $(\bar{x}, \bar{y})$ of a stable focus type. At the point $k_0{}^2$ the first catastrophe takes place: the imaginary part of the eigenvalues vanishes, both the eigenvalues in this sensitive state being equal, $\varLambda_1 = \varLambda_2$. Such a catastrophe is not related to a loss of stability by the point $(\bar{x}, \bar{y})$; only a global phase portrait changes (but not in the vicinity of this point). Next, at a further increase in k^2, the second sensitive state, k_-^2, is achieved in which one of the eigenvalues becomes zero. The spatially homogeneous stationary state $(\bar{x}, \bar{y})$ loses stability and for $k^2 > k_-^2$ dissipative structures, periodical spatial structures of type (5.121), may appear. As in this case $\mathrm{Re}(\Lambda_1) > 0$, the time evolution of solutions of this type should thus be examined using an exact non-linear equation (5.116).

The sensitive state derives from the requirement $B = 0$. The critical values, k_-^2, k_+^2 are thus equal, see equations (5.124c), (5.125b):

$$k_\pm^2 = \left[a_{11} D_y + a_{22} D_x \pm \left[(a_{11} D_y + a_{22} D_x)^2 - 4 D_x D_y |a_{ij}| \right]^{1/2} / (2 D_x D_y) \right.$$

$$|a_{ij}| \equiv a_{11} a_{22} - a_{12} a_{21} \tag{5.127}$$

The case $\mathrm{Re}(\Lambda_1) = 0$, $\mathrm{Re}(\Lambda_2) = 0$, see Fig. 88, corresponds to a most critical situation.

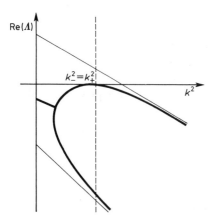

Fig. 88. Sensitive state of the dependence of Λ on k^2.

In this case $k_-^2 = k_+^2$, or

$$(a_{11} D_y + a_{22} D_x)^2 - 4 D_x D_y (a_{11} a_{22} - a_{12} a_{21}) = 0 \tag{5.128}$$

This is a sensitive state owing to the change in control parameters on which depend the elements of the stability matrix a_{ij} and the diffusion coefficients D_x, D_y.

The above considerations may be summarized as follows. The spatially homogeneous stationary state $(\bar{x}, \bar{y})$ loses stability with the generation of a spatially nonhomogeneous structure when the following conditions are satisfied owing to a suitable selection of control parameters on which a_{ij}, D_x, D_y depend:

$$a_{11}a_{22} - a_{12}a_{21} > 0 \qquad (5.129\text{a})$$

$$a_{11}D_y - a_{22}D_x > 0 \qquad (5.129\text{b})$$

$$\left(a_{11}D_y + a_{22}D_x\right)^2 > 4D_xD_y\left(a_{11}a_{22} - a_{12}a_{21}\right) \qquad (5.129\text{c})$$

$$a_{11} + a_{22} < 0 \qquad (5.129\text{d})$$

Note that it follows from inequalities (5.129b), (5.129d) that a_{11}, a_{22} have opposite signs, e.g. $a_{11} > 0$. This implies that the x variable is of an autocatalytic nature. Furthermore, since a_{11}, $|a_{22}|$ are proportional to the inverse time hierarchies associated with the x, y variables (when $a_{12} \cong 0$, $a_{21} \cong 0$, $D_x \cong 0$, $D_y \cong 0$: $x \sim \exp(a_{11}t)$, $y \sim \exp(a_{22}t)$) it follows from inequality (5.129b) that the component X should diffuse more rapidly than the component Y[X] $(x = [\text{X}], y = [\text{Y}])$.

A catastrophe corresponding to the sensitive state (5.128) is sometimes called the Turing bifurcation. Apparently, the question what is the time evolution of the solution of (5.121) type on losing stability by the spatially homogeneous stationary state $(\bar{x}, \bar{y})$ may be settled only by examining the exact equation (5.116). The methods of investigation of this problem will be discussed in the next section.

The plot $\Lambda_{1,2}(k^2)$ for the second case is shown in Fig. 89.

'n this case both the eigenvalues are real in the entire range of k^2 values. There are two sensitive states corresponding to the cases $\Lambda_1 = 0$ and $\Lambda_2 = 0$.

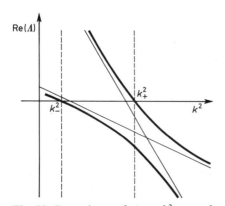

Fig. 89. Dependence of Λ on k^2, second case.

5.7.3 Classification of catastrophes associated with a loss of stability by a spatially homogeneous stationary state (wave phenomena)

We shall now show how certain wave phenomena related to the loss of stability by a spatially homogeneous stationary state in systems of equations of reactions with diffusion may be classified analogously to the case of kinetic equations of diffusionless reactions. Wave phenomena are convenient to examine by seeking solutions to the system of equations (5.116) of the travelling wave form

$$x(t, r) = x(t + \varkappa r) \equiv x(\tau) \tag{5.130a}$$

$$y(t, r) = y(t + \varkappa r) \equiv y(\tau) \tag{5.130b}$$

where we do not assume that x, y are small.

Insertion of (5.130) into (5.116) allows to write them down in the form of the following system of equations

$$x' = P(x, y) + (D_x \varkappa^2) x'' \tag{5.131a}$$

$$y' = Q(x, y) + (D_y \varkappa^2) y'' \tag{5.131b}$$

where we introduced the notation $f' \equiv df/d\tau$.

Equations (5.131) will be rearranged to an equivalent system of equations containing only the first derivatives

$$x' = u \tag{5.132a}$$

$$-D_x \varkappa^2 u' = P(x, y) - u \tag{5.132b}$$

$$y' = v \tag{5.132c}$$

$$-D_y \varkappa^2 v' = Q(x, y) - v \tag{5.132d}$$

Note that owing to chemical interpretation the inequalities $x \geqslant 0$, $y \geqslant 0$ must be satisfied while u, v can have an arbitrary sign. Linearization of (5.132)

$$x = \bar{x} + \xi, \quad u = \bar{u} + v, \quad y = \bar{y} + \eta, \quad v = \bar{v} + \zeta$$

$$|\xi|, \quad |v|, \quad |\eta|, \quad |\zeta| \ll 1 \tag{5.133}$$

where $\bar{x}$, $\bar{u}$, $\bar{y}$, $\bar{v}$ are the stationary solutions to (5.131), leads to the system of equations

$$\xi' = v \tag{5.134a}$$

$$-D_x\varkappa^2 v' = a_{11}\xi - v + a_{12}\eta \tag{5.134b}$$

$$\eta' = \zeta \tag{5.134c}$$

$$-D_y\varkappa^2\zeta' = a_{21}\xi + a_{22}\eta - \zeta \tag{5.134d}$$

The same system would be obtained by substituting (5.130) into the linearized sytem (5.120) and eliminating the second derivatives as in passing from (5.131) to (5.132).

The stability of the stationary solutions to (5.134) may be investigated using the methods previously described in Sections 5.4–5.6. Hence, substituting in (5.134) for

$$\xi = \xi_0 \exp(\Lambda t), \quad v = v_0 \exp(\Lambda t), \quad \eta = \eta_0 \exp(\Lambda t),$$

$$\zeta = \zeta_0 \exp(\Lambda t) \tag{5.135}$$

we reduce (5.134) to the form of an eigenequation

$$\Lambda\xi = v \tag{5.136a}$$

$$\Lambda v = -\left[a_{11}/(D_x\varkappa^2)\right]\xi + \left[1/(D_x\varkappa^2)\right]v - \left[a_{12}/(D_x\varkappa^2)\right]\eta \tag{5.136b}$$

$$\Lambda\eta = \zeta \tag{5.136c}$$

$$\Lambda\zeta = -\left[a_{21}/D_y\varkappa^2)\right]\xi - \left[a_{22}/D_y\varkappa^2)\right]\eta + \left[1/(D_y\varkappa^2)\right]\zeta \tag{5.136d}$$

The catastrophes involving the loss of stability by stationary solutions are related to the change in a phase portrait of the system (5.134) in the phase space $(\xi_1, \xi_2, \eta_1, \eta_2)$. The sensitive states of the systems (5.134), (5.116) are given by the requirement $\text{Re}(\Lambda) = 0$ and may be classified using the theory presented in Sections 5.4–5.6.

5.8. ANALOGIES BETWEEN ELEMENTARY AND GENERALIZED CATASTROPHE THEORY

To conclude this chapter dealing with the investigation of stability of phase portraits of dynamical systems the analogies to the methods and notions of elementary catastrophe theory will be pointed out.

Let us recall that in elementary catastrophe theory critical points of potential functions are examined. A potential function can have noncritical points, nondegenerate critical points and degenerate critical points. To degenerate critical points correspond sensitive states lying in the state variable and control parameter space in the catastrophe manifold M; their projection on the control parameter space yields the bifurcation set Σ. Catastrophes take place on crossing sensitive states of the system. When a state is sensitive with respect to a part of state variables, then by using the splitting lemma the dependence on those variables which are associated with catastrophes can be separated. By introducing the relation of equivalence of potential functions, equivalent classes of potential functions (describing catastrophes of the same type) may be classified. The most general form of a function in the class is called a universal unfolding; the number of control parameters in a universal unfolding is codimension. Eventually, universal unfoldings are proven to be structurally stable.

Analogous notions occur in catastrophe theory of dynamical systems; the procedure of their investigation is also similar. Instead of potential functions, phase portraits are examined. The points of a phase portrait may be nonsingular or nondegenerate and degenerate singular points (the latter two being stationary states). The sensitive states to which the catastrophes of a phase portrait are related correspond to degenerate stationary points. Likewise, the notions of catastrophe manifold and bifurcation set may be introduced. A counterpart to the splitting lemma is the centre manifold theorem which permits to isolate the dependence on variables associated with catastrophes. The centre manifold and Grobman–Hartman theorems enable an examination of the equivalence of phase portraits. Consequently, the classes of dynamical systems in which catastrophes of the same type occur may be classified. Standard forms correspond to universal unfoldings of elementary catastrophe theory. The number of control parameters in a standard form constitutes a codimension. The structural stability of phase portraits has been demonstrated for a number of standard forms.

Noticeably, the existing similarities and analogies are so strongly marked that one may state that for dynamical systems there is an equivalent of elementary catastrophe theory — the theory examining and classifying the phenomena of a loss of stability type (the theory is sometimes referred to as generalized catastrophe theory). It should be pointed out, however, that all possible phenomena of a catastrophe type in equations of reactions with diffusion are yet far from being understood, recognized and classified. On the other hand, the results having been obtained so far provide a rather general picture of catastrophes in dynamical systems.

Appendix to Chapter 5

A5.1 PROPERTIES OF A PHASE PLANE

Consider properties of an autonomous system

$$\dot{x} = P(x, y) \tag{A1a}$$

$$\dot{y} = Q(x, y) \tag{A1b}$$

where the functions P, Q are continuously differentiable. We will use the designation $\mathbf{F} = (P, Q)$. Let be given on the x, y phase plane a certain closed smooth line L without self-intersection points, but otherwise arbitrary. We shall consider the behaviour of points of this line with the change in time when all the points of the phase plane evolve in accordance with equations (A1).

For $t = 0$ the points x, y belonging to the line L are given by

$$x = \varphi_1(\tau), \qquad y = \varphi_2(\tau) \tag{A2}$$

and when the parameter τ varies from 0 to T the point (x, y) makes one cycle over the line L in positive direction. The new position of the line L after time t, L_t $(L_0 \equiv L)$, is expressed by

$$x = \Psi_1(t, \tau), \qquad y = \Psi_2(t, \tau) \tag{A3}$$

where Ψ_1, Ψ_2 are the solutions to (A1) satisfying the initial condition

$$\Psi_1(0, \tau) = \varphi_1(\tau), \qquad \Psi_2(0, \tau) = \varphi_2(\tau) \tag{A4}$$

It may be shown that L_t is also a smooth line without self-intersections.

The area of the figure Q_t contained by the line L_t expressed by the equation

$$|Q_t| = \int_{L_t} x\,dy = \int_0^T \Psi_1(t, \tau)\,d/d\tau\left[\Psi_2(t, \tau)\right]d\tau \tag{A5}$$

We may now compute the change in the area $|Q_t|$ with time by differentiating equation (A5) with respect to time t

$$d/dt\,|Q_t| = \int_0^T \left[(\Psi_1)_t'(\Psi_2)_\tau' + \Psi_1(\Psi_2)_{\tau t}''\right]d\tau \tag{A6a}$$

$$= \int_0^T (\Psi_1)_t'(\Psi_2)_\tau' + \int \Psi_1\left[(\Psi_2)_t'\right]_\tau'd\tau \tag{A6b}$$

Integrating by parts the second integral in (A6b) and employing the equality $(\Psi_1(\Psi_2)_t')^T_{t=0} = 0$ we obtain

$$d/dt|Q_t| = \int_0^T \left[(\Psi_1)_t'(\Psi_2)_\tau' - (\Psi_1)_\tau'(\Psi_2)_t' \right] d\tau \tag{A7}$$

Using equations (A1) we write (A7) in the form

$$d/dt|Q_t| = \int_0^T (P \partial Q/\partial\tau - Q \partial P/\partial\tau) d\tau = \int_{L_t} P dy - Q dx \tag{A8}$$

Applying now to (A8) the Green theorem we finally arrive at

$$d/d|Q_t| = \iint_{Q_t} (\partial P/\partial x + \partial Q/\partial y) dx dy \equiv \iint_{Q_t} \operatorname{div} \mathbf{F} dx dy \tag{A9}$$

where $\mathbf{F} = (P, Q)$.

Two vital conclusions follow from (A9).

1. Firstly, if in a certain connected region $\operatorname{div} \mathbf{F}$ is everywhere different from zero ($\operatorname{div} \mathbf{F} > 0$ or $\operatorname{div} \mathbf{F} < 0$) then the system (A1) cannot have limit cycles located entirely in this region. This is the so-called Bendixon criterion.

Indeed, if L were such a cycle, we would obtain $L_t = L$ (the condition for an existence of a limit cycle), that is $|Q_t| = $ const. It then follows from (A9) and the constancy of sign of $\operatorname{div} \mathbf{F}$ that $\operatorname{div} \mathbf{F} = 0$, i.e. we obtain a contradiction. Accordingly, such a cycle does not exist.

For example, the canonical system

$$\dot{x} = -\partial H/\partial y \tag{A10a}$$

$$\dot{y} = +\partial H/\partial x \tag{A10b}$$

where H is a Hamiltonian function, cannot have limit cycles because $\operatorname{div} \mathbf{F} \equiv 0$. Likewise, a linear system, i.e. such a system in which the functions P, Q in equations (A1) are linear functions of x, y, cannot have limit cycles since then $\operatorname{div} \mathbf{F} \equiv $ const. In other words, limit cycles (concentration oscillations) in a chemical reaction are possible only in the case of non-linear kinetics.

2. Secondly, for any closed line L the area $|Q_t|$ for the canonical system (A10) is constant. Indeed, substituting in the integrand in (A9) for $P = -\partial H/\partial y$, $Q = +\partial H/\partial x$ we obtain $d/dt|Q_t| = 0$. This result is known as the Liouville theorem. A consequence of the Liouville theorem is non--existence of stationary points of a node or focus type for a canonical

system: both in the case of a node and a focus (stable or unstable) the neighbourhood of a stationary point changes the area (decreases or increases). In other words, the system (A10) does not have stationary points.

We will now show that also the gradient system

$$\dot{x} = -\partial V/\partial x \tag{A11a}$$

$$\dot{y} = -\partial V/\partial y \tag{A11b}$$

where V is a potential function, cannot have limit cycles. Let $\mathbf{X}(t) = [x(t), y(t)]$ be a periodical solution of a period T. Since the following integral

$$\int\limits_{t}^{t+T} (\mathrm{d}\mathbf{X}/\mathrm{d}t')^2 \, \mathrm{d}t' = - \int\limits_{t}^{t+T} (\mathrm{d}\mathbf{X}/\mathrm{d}t') \nabla V(\mathbf{X}) \, \mathrm{d}t'$$

$$= \int\limits_{\mathbf{X}(t)}^{\mathbf{X}(t+T)} \nabla V(\mathbf{X}) \mathrm{d}\mathbf{X} = -[V(\mathbf{X})]_{\mathbf{X}(t)}^{\mathbf{X}(t+T)} \tag{A12}$$

is equal to zero, where equations (A11) and periodicity of the solution, $\mathbf{X}(t + T) = \mathbf{X}(t)$, were employed, the positive integrand is identically equal to zero: $(\mathrm{d}\mathbf{X}/\mathrm{d}t)^2 \equiv 0$. Consequently, also $\mathrm{d}\mathbf{X}/\mathrm{d}t$ vanishes, $\mathrm{d}\mathbf{X}/\mathrm{d}t \equiv 0$. We thus obtain a contradiction to the assumption that $\mathbf{X}(t)$ is a periodical solution (and, accordingly, it cannot be constant): the gradient system (A11) cannot have a limit cycle.

A5.2 THE POINCARÉ–BENDIXON THEOREM

If a trajectory of the autonomous system (A1) always remains within a certain confined region of the phase space for $t \to \infty$, then the following cases are possible: (1) the trajectory approaches a stationary point; (2) the trajectory approaches a limit cycle (periodicity of a trajectory) or is identical with it.

It should be stressed that the Poincaré–Bendixon theorem given above does not hold for systems of three or more autonomous equations; hence, in such systems non-periodicity of a nonstationary trajectory remaining within a confined region is possible.

A5.3 THEORY OF VECTOR FIELD ROTATION

Consider a vector field in the x, y plane

$$\mathbf{F}(x, y) = [P(x, y), \; Q(x, y)] \tag{A13}$$

The point (x_0, y_0) is called a nonsingular point of the field **F** if there exists a vicinity of the point (x_0, y_0) in which the field is defined and continuous and $\mathbf{F}(x_0, y_0) = 0$. Otherwise (x_0, y_0) is called a singular point of the field **F**.

Let be given in a plane a certain oriented line L not containing singular points of the field **F**. The rotation of the field **F** along the line L, $O_L\{\mathbf{F}\}$ is defined as an increment of the angle, divided by 2π and computed counterclockwise, which forms the vector $\mathbf{F}(x, y)$ with a specified direction when the point (x, y) crosses the line L according to its orientation.

If the line L is smooth in segments and the functions P, Q are continuously differentiable in a certain vicinity of the line, then the following equation is valid:

$$O_L\{\mathbf{F}\} = 1/2\pi \int_L \mathrm{d}\left[\mathrm{Arctan}\,(Q/P)\right] = 1/2\pi \int_L (P^2 + Q^2)^{-1}(P\,\mathrm{d}Q - Q\,\mathrm{d}P)$$

$$= 1/2\pi \int_L (P^2 + Q^2)^{-1}\left[(PQ_x' - P_x'Q)\mathrm{d}x + (PQ_y' - P_y'Q)\mathrm{d}y\right]$$

$$(A14)$$

If the line L is closed, then the vector field rotation has the following properties: (1) $O_L\{\mathbf{F}\}$ is an integer; (2) $O_L\{\mathbf{F}\}$ does not change when the line L or the field **F** are continuously deformed in such a way that the singular points of the field **F** during the deformation miss the line L; (3) $O_L\{\mathbf{F}\} = 0$ if in the region contained by the line L there are no singular points of the field **F**.

If we compute $O_L\{\mathbf{F}\}$ for a given point (x_0, y_0) in such a way that L is a closed line and in the region contained by L there are no singular points of the field **F** except, possibly, for the point (x_0, y_0), then such a quantity is called the index of the point (x_0, y_0) and denoted as $\mathrm{Ind}\{\mathbf{F}\}\,(x_0, y_0)$.

Properties of the vector field rotation may be applied to the investigation of the system of equations (A1). In this case the field **F**, in which the functions P, Q are the right-hand sides of the system (A1), is the velocity field tangent to its phase trajectories. The field is defined in the entire x, y plane and is continuous; therefore, its only singular points are those at which $\mathbf{F} = 0$, i.e. stationary points of the system which are generally isolated points).

Computing successively the indices for a nonsingular point and for the following stationary points: saddle, node, focus and centre we find that

$$\mathrm{Ind}\{\mathbf{F}\}_{\text{nonsingular point}} = 0, \quad \mathrm{Ind}\{\mathbf{F}\}_{\text{saddle}} = -1$$

$$\text{Ind}\{\mathbf{F}\}_{\text{node}} = \text{Ind}\{\mathbf{F}\}_{\text{focus}} = \text{Ind}\{\mathbf{F}\}_{\text{centre}} = 1 \tag{A15}$$

If we make a full rotation around the line L which is a limit cycle of the system (A1), then obviously the vector of the field $\mathbf{F}$ tangent to this line will rotate by 2π, i.e. $O_L\{\mathbf{F}\} = 1$. Accordingly, within each limit cycle there is at least one stationary point (more specifically, the sum of indices of stationary points contained within a limit cycle is equal to one).

Using the theory of vector rotation and the results contained in Sections A1, A2 it may be proved that a kinetic system of the form (A1), corresponding at most to bimolecular reactions, cannot have a limit cycle (that is, concentration oscillations may not appear in this system).

A5.4 TIME EVOLUTION ON THE TRAJECTORIES OF A DYNAMICAL SYSTEM

Information about the time evolution of a certain function, for example $f[x(t), y(t)]$ when $x(t)$, $y(t)$ are trajectories of a dynamical system, i.e. satisfying equations (A1), is frequently useful. The derivative of a function f with respect to time may be readily calculated provided that x, y evolve with time in accordance with equations (A1). Such a derivative is called the derivative of a function f (with respect to time) on the trajectories of a dynamical system (A1) or the (time) derivative with respect to the system (A1) and denoted $\mathring{f}$.

Using the formula for the derivative of a superposition of functions, followed by an application of equations (A1)

$$\mathring{f}(x, y) = \partial f/\partial x\, \dot{x} + \partial f/\partial y\, \dot{y} = \partial f/\partial x\, P(x, y) + \partial f/\partial y\, Q(x, y) \tag{A16}$$

we obtain the desired equation for the derivative of a function f (with respect to time) on the trajectories of a dynamical system (A1).

A5.5 THE LYAPUNOV FUNCTION METHOD

We will begin the description of the Lyapunov method of examination of the stability of a stationary point with the analysis of a straightforward example. Consider the following system of equations:

$$\dot{x} = y - x^3 \tag{A17a}$$

$$\dot{y} = -x - y^3 \tag{A17b}$$

Examine the properties of the function $W(x, y)$,

$$W(x, y) = x^2 + y^2 \tag{A18}$$

on the phase trajectories of the system (A17). For this purpose, let us compute the derivative with respect to time of the function W on the trajectories of the system (A17) using equation (A16)

$$\mathring{W}[x(t), y(t)] = -2x(y - x^3) - 2y(-x - y^3) = -2(x^4 + y^4) \leqslant 0 \tag{A19}$$

The function $W(x, y)$ is thus seen to decrease, except for the point $(0, 0)$, for increasing t.

In addition, we find that the function W is non-negative,

$$W(x, y) = x^2 + y^2 \geqslant 0 \tag{A20}$$

It follows from inequalities (A19), (A20) that

$$W(x, y) \to 0 \quad \text{when} \quad t \to \infty \tag{A21}$$

Furthermore, from property (A21) follows that

$$x(t) \to 0, \quad y(t) \to 0 \quad \text{when} \quad t \to \infty \tag{A22}$$

Hence, $(0, 0)$ is the point which is approached by the phase trajectories of the system (A17). The point $(0, 0)$ satisfies the equations $\dot{x} = \dot{y} = 0$ and thus is a stationary point of the system (A17).

The analysis of properties of the function $W(x, y)$ has led us to a conclusion that $(0, 0)$ is an (asymptotically) stable stationary point. Note that to arrive at this conclusion it was not necessary to solve the system (A17).

Considerations of this type are characteristic of the Lyapunov method of examination of the stability of a stationary point; the function W defined by equation (A18) is an example of the so-called Lyapunov function. Note also that to draw the conclusion about an asymptotic stability of the stationary point $(0, 0)$, in addition to inequality (A19) deriving from properties of the Lyapunov function and properties of the investigated system, inequality (A20) was also required.

A general formulation of the Lyapunov theorem concerning the stability of stationary points of autonomous systems may now be given. Let the system

$$\dot{x} = \mathbf{F}(\mathbf{x}), \quad \mathbf{F} = (F_1, ..., F_n), \quad \mathbf{x} = (x_1, ..., x_n) \tag{A23}$$

have a stationary point, for example at the origin. If there exists a function $W(\mathbf{x})$, called the Lyapunov function, such that in a certain neighbourhood of the examined point $(0, ..., 0)$ the function W has the following properties:

$$W(\mathbf{x}) \geqslant 0, \quad W(0) = 0, \quad \overset{\circ}{W}(\mathbf{x}) < 0 \tag{A24}$$

$$W(\mathbf{x}) \geqslant 0, \quad W(0) = 0, \quad \overset{\circ}{W}(\mathbf{x}) \leqslant 0 \tag{A25}$$

where $\overset{\circ}{W}$ is the derivative with respect to time of the function W computed on the trajectories of the system (A23):

$$\overset{\circ}{W}(\mathbf{x}) = \sum_i \partial W/\partial x_i \dot{x}_i = \sum_i \partial W/\partial x_i F_i \tag{A26}$$

see equation (A16), then the stationary point $(0, ..., 0)$ is stable in the case (A24) or asymptotically stable in the case (A25). These conclusions are derived from properties (A24) or (A25) in a way similar to that used for arriving at conclusions (A21), (A22) from inequalities (A19), (A20).

A physical interpretation of the Lyapunov function will now be discussed. In the case of the gradient system

$$\dot{x} = -\partial V/\partial x \tag{A27a}$$

$$\dot{y} = -\partial V/\partial y \tag{A27b}$$

the derivative of a function V with respect to time computed on the trajectories of the system (A27) according to equation (A16)

$$\overset{\circ}{V}(x, y) = -(\partial V/\partial x)^2 - (\partial V/\partial y)^2 \tag{A28}$$

is non-positive. Hence, if the potential V in a certain vicinity of a stationary point is positive, it is then the Lyapunov function.

On the other hand, in the case of a Hamiltonian system

$$\dot{x} = +\partial H/\partial y \tag{A28a}$$

$$\dot{y} = -\partial H/\partial x \tag{A28b}$$

the derivative of the function H with respect to time computed on the trajectories of the system (A28) using equation (A16)

$$\overset{\circ}{H} = (\partial H/\partial x)(\partial H/\partial y) + (\partial H/\partial y)(-\partial H/\partial x) = 0 \tag{A29}$$

is equal to zero. Hence, if a Hamiltonian function in a certain neighbourhood of a stationary point is non-negative, it is then the Lyapunov function. It may thus be stated that the Lyapunov function is a generalization of the notion of a potential function and of a Hamiltonian function.

Finally, we will give the form of the Lyapunov function for a system of two linear equations

$$\dot{x} = a_{11}x + a_{12}y \tag{A30a}$$

$$\dot{y} = a_{21}x + a_{22}y \tag{A30b}$$

and discuss some cases of stable and unstable stationary points of the system (A30). The function $W(x, y)$

$$W(x, y) = 1/2\left(\alpha x^2 + 2\beta xy + \gamma y^2\right) \tag{A31}$$

$$\alpha = \left(\Delta + a_{21}{}^2 + a_{22}{}^2\right), \quad \beta = -\left(a_{11}a_{21} + a_{12}a_{22}\right)$$

$$\gamma = \left(\Delta + a_{11}{}^2 + a_{12}{}^2\right) \tag{A32}$$

$$A = a_{11} + a_{22} > 0, \quad \Delta = a_{11}a_{22} - a_{12}a_{21} < 0 \tag{A33}$$

is the Lyapunov function of the system (A30), for it is positively definite by virtue of conditions (A33) $\left(\beta^2 - \alpha\gamma < 0, \alpha > 0, \gamma > 0\right)$ and its derivative with respect to time on the trajectories of the system (A30) is negative

$$\dot{W} = \left(\alpha x + \beta y\right)\left(a_{11}x + a_{12}y\right) + \left(\beta x + \gamma y\right)\left(a_{21}x + a_{22}\right)$$

$$= A\Delta\left(x^2 + y^2\right) < 0 \tag{A34}$$

The conditions (A33) may be shown to include cases (a1), (a2), (a3), (d) from Section 5.1. The existence of the Lyapunove function implies the stability of stationary points of this type.

A5.6 THE ISOCLINE METHOD OF CONSTRUCTING A PHASE PORTRAIT

We will give an approximate method of drawing phase portraits, that is finding the forms of solutions $\left[x(t), y(t)\right]$ on the phase plane for the system (5.2). For the purpose, the system (5.2) will be written in a different form

$$y' = f\left[x, y(x)\right], \quad f(x, y) \equiv Q(x, y)/P(x, y) \tag{A35}$$

where $y' \equiv \mathrm{d}y/\mathrm{d}x$. Equation (A35) is known to have a solution for any initial conditions, $x_0 = x(t_0)$, $y_0 = y(t_0)$, except the points at which the function $f(x, y) = \infty$ (i.e. when $P = 0$ and $Q = 0$).

The following geometric interpretation of the solution of equation (A35) may be provided: if we draw at each point (x, y) of the phase plane a section of a straight line of the slope equal to $f(x,y)$, and if the trajectory of the solution $\left[x(t), y(t)\right]$ crosses this point, then the trajectory is tangent to the section plotted at this point.

The above remark forms the basis for an approximate method of constructing the phase portraits for equation (A35), called the isocline method. At the points (x, y) where the slope of the trajectory is α, the equality $\tan(\alpha) = f(x, y)$ is fulfilled, since $f[x, y(x)] = y'(x)$. All such points lie on a line called the isocline, defined by the formula $\tan(\alpha) = f(x, y)$: isoclines connect the points at which the sections tangent to the trajectory are identically inclined with respect to the x-axis.

The main isoclines: $\tan(\alpha) = 0$, i.e. $f(x, y) = 0$ — this is the isocline of horizontal tangents, and $\tan(\alpha) = \infty$, i.e. $f(x, y) = \infty$ — this is the isocline of vertical tangents, are of special significance. The isocline $f(x, y) = 0$ divides the phase plane into the regions of identical monotonicity of solutions: in the region where $f(x, y) > 0$ the solutions increase ($y' > 0$); on the other hand, the solutions decrease in the region in which $f(x, y) < 0$ ($y' < 0$).

Let us draw, for example, an approximate phase portrait for the system of equations

$$\dot{x} = x \equiv P(x, y) \tag{A36a}$$

$$\dot{y} = -x + x^3 + xy \equiv Q(x, y) \tag{A36b}$$

The system (A36) is written in the form (A35)

$$y' = Q(x, y)/P(x, y) = -1 + x^2 + y \equiv f(x, y) \tag{A37}$$

Hence, the isoclines are the curves meeting the equation $f(x, y) = \tan(\alpha) = -1 + x^2 + y$. The isoclines of equation (A37) constitute a family of the parabolas

$$y(x) = -x^2 + 1 + \tan(\alpha) \tag{A38}$$

For the angle $\alpha = -\pi/4$ we have the equation $y = -x^2$; the isoclines of horizontal tangents, $\alpha = 0$, are given by the equation $y = -x^2 + 1$; for $\alpha = \pi/4$ we obtain the equation $y = -x^2 + 2$, etc. The isoclines for $\alpha = -\pi/4, 0, \pi/4$ and a plot of the solution to equation (A38), given by the equation

$$y(x) = Ce^x - (x + 1)^2 \tag{A39}$$

and meeting the initial condition $y(0) = -1$ $(C = 0)$, are shown in Fig. 90.

The lines dividing the phase plane into regions of a different course of trajectories are called separatrices. For example, the isocline of horizontal tangents, $f(x, y) = 0$, divides the phase plane into regions in which the solutions $y(x)$ either increase or decrease; it thus is a separatrix. We shall

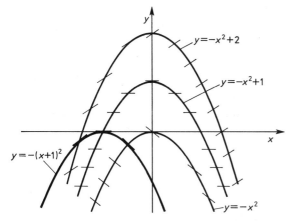

Fig. 90. The isocline method for equation (A36).

give the method of finding separatrices for linearized equations (5.6). The separatrix equation, in polar coordinates r, φ, is of the form

$$\xi(t) = r(t)\cos(\varphi), \qquad \eta(t) = r(t)\sin(\varphi) \tag{A40}$$

where φ is constant. Substitution of (A40) into (5.10) yields the equation of a separatrix (a straight line):

$$\sin(\varphi)[a_{11}\cos(\varphi) + a_{12}\sin(\varphi)] = \cos(\varphi)[a_{21}\cos(\varphi) + a_{22}\sin(\varphi)] \tag{A41}$$

A5.7 FRACTAL DIMENSION

As we have concluded in Chapter 5, in the case of the Lorenz system a trajectory always remains within a confined region of the phase space, being non-periodical. For $t \to \infty$, the trajectory approaches a certain limit set: the Lorenz attractor. It follows from the Liouville theorem that the Lorenz attractor has a zero volume (since $\mathrm{div}\,\mathbf{F} < 0$). This implies that, apparently, the Lorenz attractor is a point (dimension zero), a line (dimension one), or a plane (dimension two). Then, however, the trajectory for $t \to \infty$ would have remained within a confined region on the plane and, by virtue of the Poincaré–Bendixon theorem. Hence, a conclusion follows that the Lorenz attractor has a fractional dimension, larger than two.

From the above analysis follows the necessity of defining sets of fractional dimension. Commonly, such sets are the fractal sets (Greek fractus — broken) introduced to physics by Mandelbrot. We will confine ourselves to defining a fractal dimension, i.e. dimension of a fractal set (usually noninteger) and providing an example of the fractal set for a case of the so-called self-similar set.

Consider the following Cantor construction. A unit section is divided into three equal parts and the middle segment removed. Next, the remaining sections are divided into three equal parts and their middle segments removed again. The procedure, repeated an infinite number of times, yields the fractal Cantor set, see Fig. 91, in which the first few steps of the construction are shown.

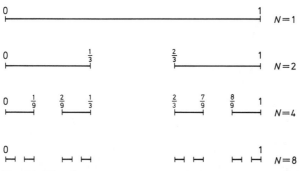

Fig. 91. The Cantor set.

In each scale (at an arbitrarily large blow-up of the figure) the Cantor set has a complex structure. This is the so-called self-similar structure: a small fragment of the Cantor set enlarged a number of times resembles the original set (more specifically, the 3^n-fold enlargement of the set preserves the image of the structure; at such an enlargement the Cantor set lying on the section $(0, 3^{-n})$ looks like the original Cantor set).

A definition of the dimension can be generalized so as to include the case of self-similar sets. The dimension d of a set embedded in the p-dimensional space $(p = 1, 2, ...)$ is defined as the limit

$$d = \lim \left\{ \ln\left[N(\varepsilon)\right] / \ln(1/\varepsilon) \right\} \tag{A42}$$

where $N(\varepsilon)$ is the number of p-dimensional cubes of side ε required to cover this set. The above definition implies that, for small ε, $N(\varepsilon)$ is given by the equation

$$N(\varepsilon) \cong \text{const } \varepsilon^{-d} \tag{A43}$$

In other words, if we want to determine the position of a set accurate to ε, $N(\varepsilon)$ cubes are required to mark the position of the set. The dimension of a set is thus related to the quantity of information required to determine its position is space.

It can be readily checked that the above definition gives a good result for regular sets. For instance, for the point $N(\varepsilon) = 1 \equiv \varepsilon^{\circ}$, the section $(0, 1)$ can be covered with sections of length ε each, if the number of these sections is equal to $N(\varepsilon) = \varepsilon^{-1}$, etc. Hence, it follows from (A43) that the dimension of such a point, section, will be 0, 1, etc., respectively.

In the case of the Cantor set we conclude, see Fig. 91, that for $\varepsilon = 1/3$, $N = 2$; for $\varepsilon = 1/9$, $N = 4, ...$; for $\varepsilon = 3^{-q}$, $N = 2^q$. Hence, it follows from (A43) that the dimension of the Cantor set is

$$N(\varepsilon) = \lim_{q \to \infty} \left[\ln(2^q)/\ln(3^{-q})\right] = \ln(2)/\ln(3) \cong 0.63 \qquad \text{(A44)}$$

The Cantor set is an example of a self-similar set (fractal) of a fractional fractal dimension.

A5.8 THE ROUTH–HURWITZ CRITERION

The Routh–Hurwitz criterion decides when a given polynomial has roots with a negative real part. Such information proves useful in the analysis of stability of stationary solutions to systems of ordinary differential equations. The examined polynomial is a characteristic polynomial of the stability matrix a_{ij}

$$\det|a_{ij} - \lambda\delta_{ij}| = 0, \qquad i,j = 1, 2, ..., n \qquad \text{(A45)}$$

The characteristic polynomial can be written in the form

$$x^n + a_1 x^{n-1} + a_2 x^{n-2} + ... + a_{n-1}x + a_n = 0 \qquad \text{(A46)}$$

From the coefficients of equation (A46), $1, a_1, ..., a_n$, we construct the so-called Hurwitz matrices

$$\Delta_n = \begin{vmatrix} a_1 & 1 & 0 & 0 & 0 & 0 & \cdots & 0 \\ a_3 & a_2 & a_1 & 1 & 0 & 0 & \cdots & 0 \\ a_5 & a_4 & a_3 & a_2 & a_1 & 1 & \cdots & 0 \\ & & & & \cdots & & & \\ 0 & 0 & 0 & 0 & 0 & 0 & & a_n \end{vmatrix} \qquad \text{(A47)}$$

The Routh–Hurwitz criterion states that the characteristic polynomial (A45) has the roots with negative real parts if and only if (iff) the following requirements:

$$\det(\Delta_1) \equiv a_1 > 0, \quad \det(\Delta_2) > 0, ..., \det(\Delta_n) > 0 \qquad \text{(A48)}$$

are fulfilled.

Bibliographical Remarks

The way of presentation of the material discussed in Chapter 5 is based on papers of Guckenheimer and on the ideas contained in papers by Stewart. A paper by Nicolis and an article by Othmer in a book published by Field and Burger constitute a very good supplement to these papers. A book by Arnol'd (1983), although rather difficult, provides much additional material. The elementary method of analysis of some dynamical catastrophes presented in Section 5.6 is patterned after Arnol'd's approach to the Hopf bifurcation in the van der Pol system described in his book (1975). A book by Gilmore provides basic information on catastrophes in dynamical systems. A paper by Stewart contains another proof (compared to Section 5.5) that Hopf bifurcation is an elementary catastrophy.

Fundamentals of the theory of ordinary differential equations are given in books by Arnol'd (1978), as well as Arrowsmith and Place. One may get acquainted with the Lorenz system in Gilmore's book. A book by Arnol'd contains more advanced information on ordinary differential equations.

A paper by Feinn and Ortoleva deals with the application of the Tikhonov theorem to chemical systems, similarly bo books by Romanovskii, Stepanova and Chernavskii, which additionally provides information on the more general Shoshitishvili theorem (covering essentially a part of Guckenheimer's results). Information on an application of the centre manifold theorem can be found in a book by Carr or by Guckenheimer and Holmes. In the Carr book, the Tikhonov theorem is compared with the centre manifold theorem.

The methods of analysis of kinetic equations accounting for diffusion are described in a book by Murray, a paper by Guckenheimer (1984), in books by Chernavskii, Romanovskii and Stepanova and in a book by Fife.

Papers by Zeeman and a paper by Steward reveal the perspectives of a development of the generalized theory of catastrophes.

References

W. I. Arnol'd, *Ordinary Differential Equations*, MIT Press, Cambridge, 1978.

W. I. Arnol'd, *Geometrical Methods in the Theory of Ordinary Differential Equations*, Springer-
-Verlag, New York–Heidelberg–Berlin, 1983.

D. K. Arrowsmith and C. M. Place, *Ordinary Differential Equations*, Chapman and Hall,
London–New York, 1982.

J. Carr, "Applications of centre manifold theory", *Appl. Math. Series*, **35**, Springer-Verlag, New
York, 1981.

D. Feinn and P. Ortoleva, "Catastrophe and propagation in chemical reactions", *J. Chem.
Phys.*, **67**, 2119 (1977).

R. J. Field and M. Burger (Eds.), *Oscillations and Travelling Waves in Chemical Systems*, John
Wiley and Sons, New York–Chichester–Brisbane–Toronto––Singapore, 1985.

P. C. Fife, "Mathematical aspects of reacting and diffusing systems", *Lecture Notes in Biomath.*,
28, 1 (1979).

R. Gilmore, *Catastrophe Theory for Scientists and Engineers*, John Wiley and Sons, New York,
1981.

J. Guckenheimer, "Multiple bifurcation problems of codimension two", *SIAM J. Math. Anal.*,
15, 1 (1984).

J. Guckenheimer, "Multiple bifurcation problems for chemical reactors", *Physica*, **20D**, 1 (1986).

J. Guckenheimer and P. Holmes, *Nonlinear Oscillations, Dynamical Systems and Bifurcations of
Vector Fields*, Springer-Verlag, New York–Berlin–Heidelberg–Tokyo, 1983.

B. D. Hassard, N. D. Kazarinoff and Y-H. Wan, *Theory and Applications of Hopf Bifurcation*,
Cambridge University Press, Cambridge, 1981.

J. E. Marsden and M. McCracken, *The Hopf Bifurcation and its Applications*, Springer-
-Verlag, New York–Berlin–Heidelberg–Tokyo, 1976.

M. Medved, "The unfolding of vector fields in the plane with a singularity of codimension 3",
Czech. Math. J., **35**, 1 (1985).

J. D. Murray, *Lectures on Nonlinear-Differential-Equation Models in Biology*, Clarendon Press,
Oxford, 1977.

G. Nicolis, "Dissipative systems", *Repts. Progr. Phys.*, **49**, 873 (1986).

J. M. Romanovskii, N. W. Stepanova and D. S. Chernavskii, *Mathematical Biophysics* (in
Russian), Nauka, Moskva 1984.

I. Stewart, "Beyond elementary catastrophe theory", *Mathematics and Computation*, **14**, 25
(1984).

E. C. Zeeman, "Bifurcation and catastrophe theory", *Contemporary Math.*, **9**, 207 (1982).

E. C. Zeeman, "Stability of dynamical systems", *Nonlinearity*, **1**, 115 (1988).

Chapter 6

Catastrophes in Chemical Reactions

6.1 INTRODUCTION

In this last chapter we shall describe catastrophes which can be observed in chemical reactions. This will be essentially a macroscopic description by means of chemical kinetic equations, wherein concentrations of reagents constitute state variables. A microscopic approach to the description of catastrophes in chemical reactions, employing the Schrödinger equation in which occurs the wave function for a single molecule of the reagent, will also be briefly discussed.

Chemical kinetics equations are commonly nonlinear and may represent diverse phenomena of a catastrophe type. Theoretical studies in this area fall into two groups. Purely model considerations belong to the first group. A certain sequence of elementary reactions — the reaction mechanism, permitted from the chemical standpoint (see the Korzukhin theorem, Chapter 4) is postulated, the corresponding system of kinetic equations is found and its solutions are examined. Such a procedure allows us to predict a possible behaviour of chemical systems. The second approach involves the investigation of a mechanism of a specific chemical reaction, having interesting dynamics.

In the present chapter we will consider a number of hypothetical reaction mechanisms, leading to the occurrence of oscillations of the reagent concentrations, the appearance of spatial structures or to a transition to the state of turbulent (chaotic) oscillations or turbulent spatial structures.

We will also investigate models corresponding to specific chemical systems. Most of the attention will be devoted to the largely examined Belousov–Zhabotinskii reaction (see Chapter 1 and Section 6.2).

All these models will be classified and examined from the viewpoint of the possibility of the occurrence of catastrophes in th systems described by them, using the methods of catastrophe theory. The analysis of the models will proceed approximately as follows. The sensitive state of the dynamical system describing kinetics associated with the investigated reaction mechanism will be determined in the first step. Subsequently, using knowledge of a standard form corresponding to the determined sensitive state, the general catastrophe dynamics possible in the examined system will be predicted. Finally, in the last step we will study some catastrophes in more detail (most frequently this will be the Hopf bifurcation).

In the end, we will describe very briefly the application of catastrophe theory to a description of chemical reactions in terms of individual molecules participating in the reaction. The approach will be based on some methods of quantum chemistry employing solutions to the Schrödinger equation. The description of chemical reactions involving a simultaneous application of the microscopic description, the Schrödinger equation, and of elementary catastrophe theory is notionally similar to the description of diffraction catastrophes for the Schrödinger equation, see Section 3.4.3.

6.2 THE BELOUSOV–ZHABOTINSKII REACTION

6.2.1 Introduction

The Belousov–Zhabotinskii (BZ) reaction has been selected as an example illustrating diverse dynamical states observable in chemical systems. The BZ reagent is very convenient both for experimental and theoretical investigations, since the BZ reaction has many dynamical states of interest, which will be described below. In the BZ reaction one may observe the steady state, the time periodic state (concentration oscillations), the spatially periodic state, the stationary state (dissipative structures), the time and spatially periodic state (propagating chemical waves) and turbulent states (chaotic oscillations, stochastic spatial structures, stochastic chemical waves).

The BZ reaction, discovered by Belousov in 1950, was met with skepticism and incredulity. In the reaction, involving oxidation of citric acid by bromate in an acidic medium in the presence of Br^- and $Ce(IV)/Ce(III)$ catalyst, Belousov has observed oscillations of concentrations of some components $(Br^-, Ce(III), Ce(IV))$ despite homogenization of the reaction mixture. The disbelief which the Belousov's results have met with is surprising, for similar observations had already been made in the history of chemistry. For example, in 1916 Morgan observed a pulsatory nature of the evolution of CO from an aqueous solution of formic acid. Several years later Bray examined another decomposition reaction — decomposition of H_2O_2 in the presence of IO_3^- and I_2. In both cases the reaction mixture was homogenized. Earlier still, in 1905, Liesegang observed the formation of periodic spatial structures upon precipitation of some precipitates (e.g. Ag_2CrO_4) from a homogeneous solution. The first report on oscillations in a chemical system already appeared in 1828. It was a paper by Fechner reporting the oscillations of current in a galvanic cell.

The observations of this type should not surprise us, because periodic phenomena frequently occur in nature, beginning with physics and ending with biology (e.g. the heartbeat, brain waves). The attempts at elucidating the nature of periodic phenomena observed in living organism should be based on the mechanisms of biochemical reactions (enzymatic reactions) rather than on physical analogies, such as pendulum oscillations. However, for a very long time chemists have been skeptical or even hostile to the notion of oscillations in homogeneous chemical systems. The above spatial or time periodicities observed in chemical systems have been attributed to nonhomogeneities (dust particles, bubbles of evolving gas, insufficient stirring of solution), believing that in truly homogeneous systems such phenomena cannot occur. This opinion has derived from a paradigm very deeply rooted on chemical ground, which can be formulated as follows: a chemical system approaches the state of equilibrium (steady state) monotonically, because this results from the thermodynamical requirement of a decrease in the free energy of the system as a whole.

It should be emphasized that there have been exceptions to this attitude. In 1910 and 1920 Lotka published his theory of chemical reactions in which the oscillations of reagent concentrations could appear. An essential feature of the Lotka models was nonlinearity. In mathematics and physics a trend has long persisted to examine linear systems and phenomena and to replace non-linear models by (approximate) linear models. The trend, originating from insufficient mathematical means, has turned into specific philosophy. The non-linear Lotka models thus constituted a deviation from a canon. Hence, general arguments of thermodynamic nature, lack of interest in non-linear models and commonness of observations of a monotonic attainment of the equilibrium in chemical reactions were the reasons for skepticism and disbelief which the results of Belousov have met with.

The situation has begun to change on obtaining new results in thermodynamics by Prigogine and his school for the systems far from the state of thermodynamic equilibrium. In addition, two crucial experimental discoveries have taken place. The Belousov system has been modified by Zhabotinskii (the starting of research on the Belousov system by Zhabotinskii was in fact the turning-point; until then all articles and reports on oscillating reactions were isolated, single papers not continued by other researchers). The new system of Belousov–Zhabotinskii is very simple to obtain under laboratory conditions and to demonstrate oscillations: the application of a suitable indicator has enabled a very impressive visualization of oscillations. Moreover, during the same period oscillations of NADH fluorescence in yeast extract in a homogeneous medium were discovered which indicated the oscillatory nature of the glycolysis reaction. A further progress in this area of chemical oscillations has become possible due to development of the theory of non-linear differential equations.

The progress in the field of investigation of oscillatory reactions has also resulted from the appearance of a new research technique, deriving from the philosophy of the Prigogine school of examination of the systems far from the equilibrium. The device which enables carrying out such investigations in chemistry is a flow reactor with stirring (the methodology of this type is more convenient than performing a reaction with a large excess of some reagents). For example, according to the second law of thermodynamics the appearance of ordering under homogeneous conditions may take place only in an open system, far from the state of thermodynamic equilibrium. Nonhomogeneous structures may appear on satisfying the thermodynamic requirement of a monotonic decrease in the free energy of the system as a whole, since they appear as a result of some side reactions. When investigating some hypothetical models of chemical reactions, Prigogine and his coworkers have indicated the possibility of formation of spatial structures (these are so-called dissipative structures) under homogeneous conditions. It should be pointed out that the first models of this type have been proposed and examined by Turing in search for a chemical model of morphogenesis. The spatial structures predicted by Turing and by Prigogine and his coworkers are formed under homogeneous conditions in the Belousov–Zhabotinskii reaction in a flow reactor with stirring.

In recent years a number of other chemical reactions exhibiting complex dynamics, including the oscillations of concentration of certain components, have been discovered. In addition to the already classical Belousov–Zhabotinskii reaction (oxidation of malonic acid by bromate in the presence of the Ce(IV)/Ce(III) catalyst), Bray–Liebhafsky reaction (decomposition of H_2O_2 in the presence of IO_3^-/I_2), similar to the preceding one Briggs–Rauscher reaction (the system consists of KIO_3, H_2O_2, $CH_2(COOH)_2$, $MnSO_4$, $HClO_4$ and starch), a series of oscillating homogeneous systems, usually containing iodine, bromine and chlorine compounds have been obtained (but also oscillatory systems in which halogen atoms did not occur). The systems reveal a variety of a typical dynamical states, such as oscillations, bistability and tristability, spatial structures, chemical waves, chaotic (turbulent) states. Experimental results have excluded nonhomogeneities of the reacting system as a reason for the formation of time (oscillations, waves) or spatial structures.

It has also been established that on changing the initial conditions the system may pass from the stationary state to any of the states mentioned above; transitions between the other states of the system are also allowed. The behaviour of this type can be described in terms of the notions of catastrophe theory. A reaction mixture is the system whose state may be represented by a number of state variables, whereas other variables, called control parameters, are varied continuously and a change in the state of the system is examined. An abrupt change in the state of the system, a catastrophe, occurs for some values of control parameters. For instance, at a continuous change in concentration of one of components the system may pass from the stationary state to the oscillatory state.

6.2.2 Dynamical states of the Belousov–Zhabotinskii reaction

6.2.2.1 Concentration oscillations (time periodicity)

Oscillations of concentrations of some intermediates in the BZ reaction occur in a rather wide range of initial concentrations of the reagents. The reaction may be carried out in a closed system, for example in a graduated cylinder provided with a stirrer, or in an open system − in a flow reactor with stirring.

Table 6.1 lists the ranges of initial concentrations of the reagents for which the concentration oscillations can be observed.

TABLE 6.1

Reagent	Concentration range (M)	Exemplary (M)	Amount
$Ce(NH_4)_2(NO_3)_6$	0.0001–0.01	0.002	0.175 g
$CH_2(COOH)_2$	0.125–0.50	0.275	4.292 g
$NaBrO_3$	0.03–0.625	0.625	1.415 g
H_2SO_4	0.5–2.5	1.5	150 ml
Ferroin	0.0006	0.0006	

Some of the reagents may be replaced with others. A different cerium salt, e.g. $Ce(NH_4)_4(SO_4)_4$, can be used, $NaBrO_3$ may be replaced with $KBrO_3$, $Mn(III)/Mn(II)$ may be employed instead of $Ce(IV)/Ce(III)$, malonic acid may be replaced with citric acid.

Ferroin is a redox indicator visualizing changes in concentration of the $Ce(IV)/Ce(III)$ system, rather stable under reaction conditions; blue colour corresponds to an excess of the $Fe(III)$ ions, red colour corresponds to an excess of the $Fe(II)$ ions. Proportions of the ferroin components are given in Table 6.2.

TABLE 6.2

Ferroin (0.025 M)

Component	Amount
$FeSO_4 \cdot 7H_2O$	0.695 g
o-Phenanthroline $\cdot$ H_2O	1.625 g
H_2O	100 ml

The procedure for generation of oscillations in the Belousov–Zhabotinskii reagent is as follows. Malonic acid and the cerium salt are dissolved in sulphuric acid in a cylinder with magnetic stirring. The solution, initially yellow, becomes, colourless after several minutes. $NaBrO_3$ is then added with vigorous stirring of the reaction mixture. After approximately one minute we begin to observe periodic colouring (to yellow) and decolouring of the solution. Upon addition of ferroin the colour of the solution will alternate between blue and red.

Yellow (blue) colour corresponds to an excess of the $Ce(IV)$ ions while the colourless (red) solution corresponds to an excess of the $Ce(III)$ ions. Changes in concentrations may be followed directly or measured potentiometrically $\left(Br^-, Ce(IV)/Ce(III)\right)$ or colorimetrically (without an addition of ferroin) — $Ce(IV)$ absorbs radiation of a wavelength about 340 nm. The observations, and particularly the quantitative measurements, allows us to distinguish four fundamental phases of oscillations of concentrations, see Fig. 92.

The red solution rapidly turns to blue and then slowly becomes violet. The violet colour slowly passes into red and, after some time, a rapid change in colour to blue takes place, the colour changing in part of the cylinder, e.g. at its bottom, and then rapidly spreading throughout the stirred solution.

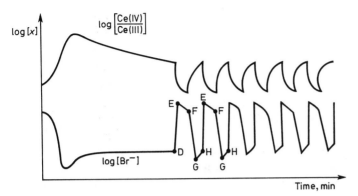

Fig. 92. Concentration oscillations in the Belousov–Zhabotinskii reaction. Reprinted with permission from: R.J. Field, E. Körös and R.M. Noyes, *Journal of American Chemical Society*, **94** (1972), 8649.

After some time (of the order of 1–3 hours) the oscillations disappear and the stationary state (equilibrium) is attained.

6.2.2.2 Spatial structures (space periodicity)

The difference between this experiment and the previous one involves mainly the fact that now the solution is not stirred. Concentrations of the reactants are given in the third column of Table 6.1.

The reagents listed in Table 6.1, except the cerium salt, should be mixed in a graduated cylinder provided with a magnetic stirrer. The solution should be stirred for some time, and then the stirring is stopped. When the movement of liquid in the cylinder ceases, an aqueous solution of the cerium salt should be added using a pipette.

The aqueous solution of cerium sat, having a density lower than sulphuric acid, forms the upper layer in the system. In such a case coloured bands at the phase boundary are formed in the cylinder. If the cerium salt is added more vigorously, bands are formed throughout the solution. On stirring asynchronous oscillations appear in various parts of the solution; sometimes a blue wave travelling along the cylinder axis may be observed.

6.2.2.3 Waves

The following way of wave generation in the BZ reagent can be given: three solutions are prepared, see Table 6.3. Next, 0.5 ml of the first solution is added to 6 ml of the second solution placed in a small beaker.

TABLE 6.3

Solution	Composition
1	3 ml of concentrated H_2SO_4, 10 g $NaBrO_3$, dissolved in 134 ml of water
2	1 g NaBr dissolved in 10 ml of water
3	2 g $CH_2(COOH)_2$ dissolved in 20 ml of water

Then, 1 ml of the third solution is added and after a few minutes the resulting solution becomes colourless. One milliliter of 0.025 M ferroin solution is then added. After thorough mixing the solution is transferred onto to a Petri dish about 9 cm in diameter and covered with a watch glass when it becomes uniformly red. After a few minutes blue dots appear against the red background. The dots propagate in the form of blue rings: inside a blue circle the colour turns to red, against which appears a growing blue dot, etc., see Fig. 93.

Fig. 93. Concentric waves in the Belousov–Zhabotinskii reaction.

Spiral (coil) waves can be generated by striking lightly the Petri dish so as to break the wave front (ring); then the disrupted ends of the wave front wind around a common centre forming spiral structures (Fig. 94).

The ring waves do not appear in the filtered BZ reagent, but are generated on repeated contamination of the solution with dust. Apparently, the process of their formation is initiated on nonhomogeneities present in the solution (heterogeneity of the system). In contrast, the spiral waves

Fig. 94. Spiral waves in the Belousov–Zhabotinskii reaction.

appear even in the thoroughly filtered BZ reagent, a given type of the coil
wave always having the same characteristic irrespective of the conditions of
its generation. Hence, such waves seem not to be generated by dust particles.
Moreover, the number of nuclei (leading centres) of spiral waves formed in
the reaction was found not to depend on whether or not the Belousov–
–Zhabotinskii reagent was contaminated with dust.

6.2.2.4 Chaotic state

A chaotic state of the Belousov–Zhabotinskii reaction may be generated
in a flow reactor. Concentrations of the reagents were found to vary
randomly for the following reagent concentrations in the influent streams at
39.6°C, see Table 6.4. The changes in Ce(IV) concentration occurring during

TABLE 6.4

Solution	Concentration (M)
$Ce_2(SO_4)_3$	0.00058
$NaBrO_3$	0.0018
$CH_2(COOH)_2$	0.056
H_2SO_4	1.5

the reaction are conveniently followed colorimetrically by measuring the
absorbance at 340 nm. A control parameter, variable in the course of
examining the reaction, is a (laminar) flow rate of the reagents through
a reactor. An exemplary dependence of $[Ce^{4+}]$ on time for a chaotic state is
shown in Fig. 95. The irregular (chaotic) dependence of the Ce^{4+} concentration on time is evident.

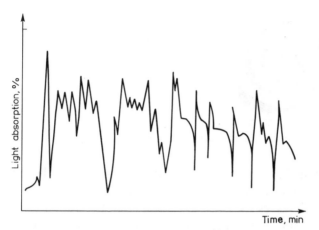

Fig. 95. Chaotic oscillations of the $[Ce^{4+}]$ concentration in the Belousov–Zhabotinskii reaction.

6.2.3 Qualitative description of mechanism of the Belousov–Zhabotinskii reaction (the Fields–Körös–Noyes mechanism)

The principal reagents are bromate BrO_3^- and malonic acid $CH_2(COOH)_2$. In the course of the reaction they are slowly consumed and their concentrations monotonically decrease. Thus, in the main reaction, oxidation of malonic acid by bromate (catalyzed by cerium ions), oscillations do not occur although the rate of the main reaction undergoes slight changes. In addition, in 1–2 M H_2SO_4 hydrogen ions are also present in large excess and the H^+ concentration is not appreciably altered during the reaction.

The reaction in which an equilibrium between BrO_3^-, Br^- and Br_2 ions is established, is a complex reaction. The mechanism of this reaction was partly discussed in Chapter 4, see equations (4.6)–(4.18).

Let us write down the elementary reaction (4.9) once again:

$$BrO_3^- + Br^- + 2H^+ \rightarrow HBrO_2 + HOBr \qquad (6.1a)$$

$$HBrO_2 + Br^- + H^+ \rightarrow 2HOBr \qquad (6.1b)$$

$$HOBr + Br^- + H^+ \rightleftharpoons Br_2 + H_2O \qquad (6.1c)$$

The presence of Br_2 in solution cannot be detected, since bromine rapidly reacts with malonic acid

$$Br_2 + CH_2(COOH)_2 \rightarrow BrCH(COOH)_2 + Br^- + H^+ \qquad (6.2a)$$

$$Br_2 + BrCH(COOH)_2 \rightarrow Br_2C(COOH)_2 + Br^- + H^+ \qquad (6.2b)$$

with the formation of brominated organic compounds. Cerium (IV) ion oxidizes these compounds and malonic acid yielding Br^- and CO_2:

$$6Ce^{4+} + CH_2(COOH)_2 + 2H_2O \rightarrow 6Ce^{3+} + HCOOH +$$
$$+ 2CO_2 + 6H^+ \qquad (6.3a)$$

$$4Ce^{4+} + BrCH(COOH)_2 + 2H_2O \rightarrow 4Ce^{3+} + Br^- +$$
$$+ HCOOH + 2CO_2 + 5H^+ \qquad (6.3b)$$

As formic acid has not been detected in the solution either, it should undergo further reactions, for example oxidation by bromine:

$$Br_2 + HCOOH \rightarrow 2Br^- + CO_2 + 2H^+ \qquad (6.4)$$

In equation (6.4) regeneration of Br^- takes place; this is an essential feature of the BZ reaction mechanism mandatory for the occurrence of oscillations in the Br^- concentration (Ce^{4+}, consumed in reactions (6.3), is regenerated similarly, see equation (6.6b)).

The overall reactions of oxidation of malonic acid by bromate can be written as follows:

$$2BrO_3^- + 3CH_2(COOH)_2 + 2H^+ \rightarrow 2BrCH(COOH)_2 +$$
$$+ 3CO_2 + 4H_2O \qquad (6.5a)$$

$$2BrO_3^- + 2CH_2(COOH)_2 + 2H^+ \rightarrow Br_2CHCOOH +$$
$$+ 4CO_2 + 4H_2O \qquad (6.5b)$$

the contributions of reactions (6.5a) and (6.5b) being 3/4 and 1/4, respectively.

Additionally, BrO_3^- can be reduced by Ce^{3+}

$$BrO_3^- + HBrO_2 + H^+ \rightleftharpoons 2BrO_2 + H_2O \qquad (6.6a)$$

$$Ce^{3+} + BrO_2 + H^+ \rightarrow Ce^{4+} + HBrO_2 \qquad (6.6b)$$

which gives rise to regeneration of Ce^{4+} and oscillations of concentrations of Ce^{3+} and Ce^{4+}.

Addition of (6.6a) to two (6.6b) yields the overall reaction

$$2Ce^{3+} + BrO_3^- + HBrO_2 + 3H^+ \rightarrow 2Ce^{4+} + 2HBrO_2 + H_2O \quad (6.7)$$

We may now present the overall mechanism proposed by Fields, Körös and Noyes, consisting of eleven reaction (R1)–(R11):

R1: $HOBr + Br^- + H^+ \rightleftharpoons Br_2 + H_2O$ (6.1c)

$k_{R1} = 8 \times 10^9 \ (\text{mole}^{-2} \ l^2 \ s^{-1}), \quad k_{-R1} = 110 \ (s^{-1})$

R2: $HBrO_2 + Br^- + H^+ \rightarrow 2HOBr$ (6.1b)

$k_{R2} = 2 \times 10^9 \ (\text{mole}^{-2} \ l^2 \ s^{-1})$

R3: $BrO_3^- + Br^- + 2H^+ \rightarrow HBrO_2 + HOBr$ (6.1a)

$k_{R3} = 2.1 \ (\text{mole}^{-3} \ l^3 \ s^{-1})$

R4: $2HBrO_2 \rightarrow BrO_3^- + HOBr + H^+$

$k_{R4} = 4 \times 10^7 \ (\text{mole}^{-1} \ l \ s^{-1})$

R5: $BrO_3^- + HBrO_2 + H^+ \rightleftharpoons 2BrO_2 + H_2O$ (6.6a)

$k_{R5} = 1.0 \times 10^4 (\text{mole}^{-2} \ l^2 \ s^{-1}), \ k_{-R5} = 2 \times 10^7 (\text{mole}^{-1} \ l \ s^{-1})$

R6: $BrO_2 + Ce^{3+} + H^+ \rightarrow HBrO_2 + Ce^{4+}$ (6.6b)

rapidly, exact kinetic data are not available

R7: $BrO_2 + Ce^{4+} + H_2O \rightarrow BrO_3^- + Ce^{3+} + 2H^+$

exact kinetic data are unavailable

R8: $Br_2 + CH_2(COOH)_2 \rightarrow BrCH(COOH)_2 + Br^- + H^+$

$r_{R8} = 1.3 \times 10^{-2} \ [H^+] \ [\text{mal}]$ (6.2a)

R9: $6Ce^{4+} + CH_2(COOH)_2 + 2H_2O \rightarrow 6Ce^{3+} + HCOOH +$

                  $+ 2CO_2 + 6H^+$

$r_{R9} = 8.8 \times 10^{-2} \ [Ce^{4+}] \ [\text{mal}]/\{0.53 + [\text{mal}]\}$ (6.3a)

R10: $4Ce^{4+} + BrCH(COOH)_2 + 2H_2O \rightarrow 4Ce^{3+} + Br^- +$

                  $+ HCOOH + 2CO_2 + 5H^+$

$r_{R10} = 1.7 \times 10^{-2} \ [Ce^{4+}] \ [\text{Brmal}]^2/\{0.20 + [\text{Brmal}]\}$ (6.3b)

R11: $Br_2 + HCOOH \rightarrow 2Br^- + CO_2 + 2H^+$

$r_{R11} = 7.5 \times 10^{-3} \ [Br_2] \ [HCOOH]/[H^+]$ 6.4)

where mal $\equiv CH_2(COOH)_2$, Brmal $\equiv BrCH(COOH)_2$ and the dimension of the rate constants of the reactions $r_{R8}, ..., r_{R11}$ is mole $l^{-1} \ s^{-1}$.

The Fields–Körös–Noyes mechanism, explaining the occurrence of oscillations of Br^-, Ce^{3+}, Ce^{4+} concentrations, can be summarized in the following way:

(1) The first reaction pathway.

When the Br^- concentration is large, BrO_3^- is reduced by Br^- to Br_2 in reactions (R3), (R2), (R1):

$$BrO_3^- + 5Br^- + 6H^+ \rightarrow 3Br_2 + 3H_2O \tag{F}$$

$$(F) \equiv (R3) + (R2) + 3(R1)$$

Br_2 reacts immediately with malonic acid according to (R8). Both the reactions are represented by the overall reaction:

$$BrO_3^- + 2Br^- + 3CH_2(COOH)_2 + 3H^+ \rightarrow 3BrCH(COOH)_2 +$$
$$+ 3H_2O \tag{A}$$

$$(A) \equiv (F) + 3(R8)$$

As explained earlier, the rate of reaction (F) is limited by the first step of reaction (R3). Since reaction (R8) is rapid compared with reaction (R3), the rate of the process (A) is given by

$$r_A = r_F = r_{R3} = -d[BrO_3^-]/dt = k_{R3}[BrO_3^-][Br^{-1}][H^+]^2 \tag{6.8}$$

The Br^- concentration decreases via this reaction path.

(2) The second reaction pathway.

When $[Br^-]$ is small, BrO_3^- may react with Ce^{3+} (reduction), because the reactions of pathway 1 are incapable of competing with the reactions of pathway 2: Br^- is considerably more active towards BrO_3^- than Ce^{3+}:

$$2Ce^{3+} + BrO_3^- + HBrO_2 + 3H^+ \rightarrow 2Ce^{4+} + H_2O + 2HBrO_2 \tag{G}$$

$$(G) \equiv (R5) + 2(R6)$$

A characteristic feature of the Fields–Körös–Noyes mechanism is the postulate of two possible reaction pathways (step 1 and step 2), controlled by the Br^- concentration.

In the case of reaction (G) the rate-determining step is reaction (R5)

$$r_G = -d[BrO_3^-]/dt = +d[HBrO_2]/dt$$
$$= k_{R5}[BrO_3^-][HBrO_2][H^+] \tag{6.9}$$

Bromic acid is produced autocatalytically and initially its concentration increases exponentially. However, the increase is limited by the disproportionation reaction

$$2HBrO_2 \rightarrow HOBr + BrO_3^- + H^+ \tag{R4}$$

The overall result of reactions $2(G) + (R4)$ is given by the equation

$$BrO_3^- + 4Ce^{3+} + 5H^+ \rightarrow HOBr + 4Ce^{4+} + 2H_2O \tag{B}$$

$$(B) \equiv 2(G) + (R4)$$

When the Br_2 concentration is small, the rate of process (B) is limited by the first step of reaction (R5):

$$r_B = 1/2\, r_G = 1/2\, r_{R5} = 1/2\, k_{R5}\left[BrO_3^-\right]\left[H^+\right]\left[HBrO_2\right] \tag{6.10}$$

since for every BrO_3^- ion consumed in process (B), two BrO_3^- ions must be used up in reaction (G).

Processes (A), pathway 1, and (B), pathway 2, should be regarded as two possible ways of reacting for $HBrO_2$. When the Br^- concentration is large, $[HBrO_2]$ is maintained at a low level due to reaction (R2), while Br^- is slowly consumed in process (A). When the Br^- concentration decreases below a critical value, the autocatalytic reaction of production of $HBrO_2$ from BrO_3^-, step (R5), predominates over reaction (R2). At this stage the $HBrO_2$ concentration increases exponentially while $[Br^-]$ still decreases rapidly due to reaction (R2). In this phase of the reaction many Ce^{3+} ions are oxidized to Ce^{4+} via reaction (R6). Figure 92 shows time variations of $[Br^-]$ and $[Ce^{4+}]/[Ce^{3+}]$. As can be seen, two steps can be distinguished on each of the reaction pathways: steps EF, FG on pathway 1 and steps GH, HE on pathway 2.

The critical Br^- concentration can be determined by subtracting reactions (R2), (R5):

$$
\begin{aligned}
d[HBrO_2]/dt &\sim -k_{R2}\left[HBrO_2\right]\left[Br^-\right]\left[H^+\right] + \\
&+ k_{R5}\left[BrO_3^-\right]\left[HBrO_2\right]\left[H^+\right] \sim \\
&\sim k_{R5}\left[BrO_3^-\right] - k_{R2}\left[Br^-\right]\left[HBrO_2\right] \sim \\
&\sim \begin{cases} -[HBrO_2] \ \text{when}\ k_{R5}\left[BrO_3^-\right] < k_{R2}\left[Br^-\right] \\ +[HBrO_2] \ \text{when}\ k_{R5}\left[BrO_3^-\right] > k_{R2}\left[Br^-\right] \end{cases}
\end{aligned}
\tag{6.11}
$$

Hence, the critical $[Br^-]$ value is obtained from the requirement $d[HBrO_2]/dt = 0$:

$$k_{R5}\left[BrO_3^-\right] = k_{R2}\left[Br^-\right]_{cr} \tag{6.12a}$$

$$\left[Br^-\right]_{cr} = \left(k_{R5}/k_{R2}\right)\left[BrO_3^-\right] = 5 \times 10^{-6}\left[BrO_3^-\right] \tag{6.12b}$$

Br^- is regenerated in the reaction of oxidation of brominated organic compounds by Ce^{4+}. For example, summation of equations (R10) + + (R11) + (R1) yields the process

$$HOBr + 4Ce^{4+} + BrCH(COOH)_2 + H_2O \rightarrow 2Br^- +$$
$$+ 4C^{3+} + 3CO_2 + 6H^+ \tag{C}$$

$$C \equiv (R10) + (R11) + (R1)$$

in which Ce^{4+} is reduced to Ce^{3+} and the Br^- concentration increases. The critical Br^- concentration at which the BZ reaction switches from pathway (B) to pathway (A) is smaller than $[Br^-]_{cr}$ owing to the fact that the ratio $[Ce^{4+}]/[Ce^{3+}]$ is much larger than $[Br^-]$.

Processes (A), (B), (C) compose the overall reaction

$$5Br_3^- + 6CH_2(COOH)_2 + 8Ce^{3+} + 2Br^- + 15H^+ \rightarrow 2HOBr +$$
$$+ 5BrCH(COOH)_2 + CO_2 + 8Ce^{4+} + 11H_2O \tag{D}$$

$$(D) \equiv 2(A) + 3(B) + (C)$$

Apparently, at certain initial concentrations the competing processes (A), (B) may monotonically approach the steady state. Such a behaviour is even observed during the initial period of the BZ reaction, prior to the occurrence of oscillations, see Fig. 92.

6.2.4 Simplified model of mechanism of the BZ reaction — the Oregonator

On the basis of the preceding discussion the five most important reactions in the Fields–Körös–Noyes mechanism can be indicated. These are the following reactions:

$$BrO_3^- + Br^- + 2H^+ \rightarrow HBrO_2 + HOBr \tag{R3}$$

$$HBrO_2 + Br^- + H^+ \rightarrow 2HOBr \tag{R2}$$

$$2Ce^{3+} + BrO_3^- + 3H^+ \rightarrow 2Ce^{4+} + H_2O + 2HBrO_2 \tag{G}$$

$$2HBrO_2 \rightarrow BrO_3^- + HOBr + H^+ \tag{R4}$$

$$4Ce^{4+} + BrCH(COOH)_2 + 2H_2O \rightarrow 4Ce^{3+} + Br^- +$$
$$+ HCOOH + 2CO_2 + 5H^+ \tag{R10}$$

Step (R3) determines the rate of the reaction via pathway (A). Step (R2) is crucial for the control of switching from pathway (A) to pathway (B). In reaction (G) $HBrO_2$ is autocatalytically produced via route (B). This is not an elementary reaction: the rates of reaction (G) and process (B) are limited by step (R5)

$$BrO_3^- + HBrO_2 + H^+ \rightleftharpoons 2BrO_2 + H_2O \tag{R5}$$

Step (R4) determines the increase in $HBrO_2$ concentration. Step (R10) initiates regeneration of Br^- from brominated organic substances in process (C).

Fields and Noyes have proposed the following mechanism, called Oregonator, based on the reactions given above:

$$A + Y \xrightarrow{k_1} X + P \tag{6.13a}$$

$$X + Y \xrightarrow{k_2} 2P \tag{6.13b}$$

$$A + X \xrightarrow{k_3} 2X + 2Z \tag{6.13c}$$

$$2X \xrightarrow{k_4} A + P \tag{6.13d}$$

$$Z \xrightarrow{k_5} f Y \tag{6.13e}$$

The following denotations were introduced in equations (6.13): $A \equiv BrO_3^-$, $P \equiv HOBr$, $Y \equiv Br^-$, $Z \equiv Ce^{4+}$. Steps (6.13a), (6.13b), (6.13d) correspond to reactions (R3), (R2), (R4); step (6.13c) has the stoichiometry of step (G) and the kinetics of reaction (R5); step (6.13e) corresponds to regeneration of Br^- at the expense of Ce^{4+} in reaction (R10); f is an empirical numerical parameter; at this stage of our considerations f is indeterminate. In the original Fields–Noyes scheme $Z \equiv 2Ce^{4+}$.

The order of magnitude of the constants $k_1, ..., k_5$ is given in Table 6.5.

TABLE 6.5

Rate constants $k_1, ..., k_s$ $(mole^{-1}s^{-1})$

$k_1 = k_{R3}[H^+]^2$	$\cong 2.1$	
$k_2 = k_{R2}[H^+]$	$\cong 2 \times 10^9$	
$k_3 = k_{R5}[H^+]$	$\cong 10^4$	
$k_4 = k_{R4}$	$\cong 4 \times 10^7$	
k_5	$\cong 0.4 [BrMal],$	$[BrMal] \ll 2 \times 10^{-1}$ (mole)

Fields and Noyes considered the case of the Belousov–Zhabotinskii reaction proceeding in well stirred solution, at a constant temperature and pressure, in a closed system. They further assumed that the changes in $[A] = [BrO_3^-]$ may be disregarded in a first approximation, that the product $P = HOBr$ has no effect on the reaction kinetics, and that the reversibility of the above reaction can be neglected.

6.3 KINETIC MODELS AND THEIR ANALYSIS

6.3.1 Kinetic equations of reactions without diffusion

6.3.1.1 Introduction

In the following sections a number of mechanisms corresponding both to real and to hypothetical reactions will be described and the respective kinetic equations given. The sequence of models to be discussed corresponds essentially to their increasing complexity. A part of the models under discussion have been examined in a similar sequence by Ebeling.

In principle it can be stated that kinetic equations containing one state variable may represent stationary states, autocatalytic processes and the phenomena related to multistability. Equations in two variables can describe periodical oscillations with time and periodical spatial structures (accounting for diffusion). Equations in three variables enable a description of chaotic processes. State variables are the reagent concentrations and, when diffusion is taken into account, additional state variables associated with the wavelengths of waves propagating in solution appear.

The models selected below permit us to represent all the phenomena mentioned above or, more specifically, to describe the transition of a system from a stable stationary state, by way of its destabilization, to a new stable state of a different type of dynamics.

The examined systems have three types of sensitive states:

(1) of the $\lambda_1 = \lambda_2$ type, related to a global change of the phase portrait, without a local change of the phase portrait in the vicinity of the examined stationary point. Sensitive states of this type usually do not lead to a distinct alteration of the dynamics of a reaction;

(2) of the $\text{Re}(\lambda_{1,2}) = 0$ type, related to a local change of a phase portrait in the vicinity of the examined stationary state. Sensitive states of this type give rise to the occurrence of space nonhomogeneities (spatial structures) or

time nonhomogeneities (oscillations) in the course of a reaction. When the nonhomogeneities are not of a periodical nature we may deal with chaotic oscillations in the case of time nonhomogeneity (observed experimentally) or with spatial structures of a stochastic type in the case of space non-homogeneities (also observed experimentally). In the case of both space and time nonhomogeneities we have to deal with propagating chemical waves;

(3) of a change in the number of stationary points type. Sensitive states of this type are experimentally reflected in the form of multistability and hysteresis phenomena.

After determining a sensitive state we will be able to predict, on the basis of the corresponding standard form, the most general dynamics possible in the investigated system and that related to catastrophes.

6.3.1.2 The reactant–product equilibrium

In this reaction a reactant A is in equilibrium with a product X:

$$A \underset{k_{-1}}{\overset{k_1}{\rightleftharpoons}} X \tag{6.14}$$

A kinetic equation, provided that [A] is maintained at a constant level, is of the form (where $x = [X]$, $a = [A]$):

$$\dot{x} = k_1 a - k_{-1} x \tag{6.15}$$

The stationary state for kinetic equation (6.15) is given by (see equation (5.4))

$$\bar{x} = k_1 a / k_{-1} \tag{6.16}$$

Equation (6.15) is linear, thus substitution (5.5) leading to the equation

$$\dot{\xi} = -k_{-1} \xi \tag{6.17}$$

allows to find an exact solution. The eigenvalue is computed from (5.7), (5.8)

$$\lambda = -k_{-1} < 0 \tag{6.18}$$

Hence, the stationary state (6.16) is always stable and in the system represented by kinetic equation (6.15) catastrophes cannot occur.

6.3.1.3 The reactant–product equilibrium in the presence of a catalyst

A mechanism of the reaction is of the form

$$A + Y \underset{k_{-1}}{\overset{k_1}{\rightleftharpoons}} X + Y \tag{6.19}$$

and the corresponding kinetic equation, provided that [A] is maintained at a constant level, looks as follows

$$\dot{x} = k_1 a y - k_{-1} y x \tag{6.20}$$

where $x = [X]$, $y = [Y]$, $a = [A]$.

The stationary state for kinetic equation (6.20) is given by

$$\bar{x} = k_1 a / k_{-1} \tag{6.21}$$

Substitution (5.5) yields an exact linear equation. The eigenvalue is computed on the basis of (5.7), (5.8)

$$\lambda = -k_{-1} y < 0 \tag{6.22}$$

The stationary state (6.21) is always stable — in the system represented by (6.20) catastrophes cannot occur.

6.3.1.4 Irreversible autocatalytic reaction

A mechanism of this reaction can be represented by the following equation

$$A + X \xrightarrow{k_1} 2X \tag{6.23}$$

and the corresponding kinetic equation, when [A] is maintained at a constant level, is of the form $(x = [X],\ a = [A])$:

$$\dot{x} = k_1 a x \tag{6.24}$$

This is not a standard kinetic system. In Section 4.3.2 we have shown how equations of this type may appear as slow dynamics of a standard kinetic system.

The stationary state for kinetic equation (6.24) is given by

$$\bar{x} = 0 \tag{6.25}$$

Substitution (5.5) yields an exact equation and the eigenvalue is computed from (5.7), (5.8)

$$\lambda = +k_1 a > 0 \tag{6.26}$$

Hence, the stationary state (6.25) is always unstable.

6.3.1.5 Reversible autocatalytic reaction

A mechanism of this reaction may be represented by the following equation:

$$A + X \underset{k_{-1}}{\overset{k_1}{\rightleftharpoons}} 2X \tag{6.27}$$

A kinetic equation, provided that [A] is maintained constant, is of the form

$$\dot{x} = k_1 a x - k_{-1} x^2 \tag{6.28}$$

where $x = [X]$, $a = [A]$.

Note that equation (6.28) has a form of the transcritical bifurcation, (5.69). Equation (5.69) may describe the catastrophe occurring on a change of sign of the parameter a. However, an analogous coefficient in equation (6.28), $k_1 a$, is always non-negative. Therefore, in the chemical system described by (6.28) a catastrophe cannot take place. The stationary state $\bar{x}_1 = 0$ is unstable and the state $\bar{x}_2 = k_1 a / k_{-1}$ is stable in the entire range of variability of the control parameters k_1, k_{-1}, a.

6.3.1.6 Reversible autocatalytic reaction with a following reaction

A mechanism of this reaction can be represented by the following equations:

$$A + X \underset{k_{-1}}{\overset{k_1}{\rightleftharpoons}} 2X \tag{6.29a}$$

$$X \xrightarrow{k_2} F \tag{6.29b}$$

The respective kinetic equation, provided that [A] is maintained at a constant level, are of the form $(x = [X], a = [A])$:

$$\dot{x} = (k_1 a - k_2) x - k_{-1} x^2 \tag{6.30}$$

Equation (6.30) has a standard form of the transcritical bifurcation, see (5.69). A transcritical catastrophe may occur since the coefficient $k_1 a - k_2$ may change sign. However, properties of the system (6.30) differ somewhat from those of a general system (5.69). Note that the stationary states of kinetic equation (6.30) are given by

$$\bar{x} = 0 \tag{6.31a}$$

$$\bar{x}_2 = (k_1 a - k_2)/k_{-1}, \quad k_1 a - k_2 \geqslant 0 \tag{6.31b}$$

The stationary state $\bar{x}_2$ has a chemical meaning only if $k_1 a \geqslant k_2$, since concentration is a non-negative quantity. On the basis of this conclusion and the analysis of the transcritical bifurcation, see Section 5.5.2.2, the occurrence of the following phenomena in the system (6.30) can be predicted.

When $k_1 a - k_2 < 0$, the system has just one stationary state, $\bar{x}_1 = 0$, which is stable. In contrast, when $k_1 a - k_2 > 0$, the system has two stationary states: $\bar{x}_1 = 0$, an unstable state, and $\bar{x}_2 = (k_1 a - k_2)/k_{-1}$, a stable state. Recall that in a standard system of the transcritical bifurcation (5.69) the catastrophe of stability exchange between the two existing states $\bar{x}_1$ and $\bar{x}_2$ takes place. On the other hand, in the chemical system (6.30) the catastrophe involves destabilization of the state $\bar{x}_1 = 0$ with a simultaneous appearance of the stable state $\bar{x}_2 > 0$.

6.3.1.7 The Schloegl reaction

A mechanism of the Schloegl reaction is as follows

$$A + 2X \underset{k_{-1}}{\overset{k_1}{\rightleftharpoons}} 3X \tag{6.32a}$$

$$X \underset{k_{-2}}{\overset{k_2}{\rightleftharpoons}} F \tag{6.32b}$$

an overall reaction having the form $A \rightleftharpoons F$. We assume that $[A]$ and $[F]$ are maintained at a constant level.

A kinetic equation is of the form

$$\dot{x} = k_1 a x^2 - k_{-1} x^3 - k_2 x + k_{-2} f \tag{6.33}$$

where $x = [X]$, $a = [A]$, $f = [F]$.

Introducing to (6.33) dimensionless variables by means of the substitution

$$t \to (k_1^2 a^2 / k_{-1}) t, \qquad x \to (k_1 a / k_{-1}) x \tag{6.34}$$

we obtain

$$\dot{x} = -x^3 + x^2 - \beta x + \gamma$$
$$\beta = k_{-1} k_2 / (k_1^2 a^2), \qquad \gamma = k_{-1}^2 k_{-2} f / (k_1^3 a^3) \tag{6.35}$$

Kinetic equation (6.35) will be rearranged to a simpler form using the substitution $x = y + 1/3$ and obtaining a standard (gradient) form of the cusp bifurcation

$$\dot{y} = -y^3 + by + a, \quad b \equiv (1/3 - \beta), \quad a \equiv (\gamma + 2/27 - 1/3\,\beta)$$

$$\beta \equiv k_{-1}k_2/(k_1{}^2 a^2) \geqslant 0, \quad \gamma \equiv k_{-1}{}^2 k_{-2} f/(k_1{}^3 a^3) \geqslant 0 \qquad (6.36)$$

The stationary states of (6.36) $\bar{y}_1$, $\bar{y}_2$, $\bar{y}_3$ satisfy the equation

$$Q(\bar{y}_i) \equiv -\bar{y}_i{}^3 + b\bar{y}_i + a = 0, \quad i = 1, 2, 3 \qquad (6.37)$$

For suitable values of the control parameters β, γ equation (6.37) has three real roots.

We will now linearize the system (6.36) in the vicinity of the stationary state $(\bar{y}_i)$, where $\bar{y}_i$ is the ith root of equation (6.37). Substituting in (6.36)

$$y \cong \bar{y}_i + \eta_i, \quad |\eta_i| \ll 1, \quad i = 1, 2, 3 \qquad (6.38)$$

we obtain the following linear equation

$$\dot{\eta} = dQ/dy\,(\bar{y}_i)\eta_i \qquad (6.39)$$

The stationary state $\bar{y}_i$ is stable when

$$\lambda = (dQ/dy)(\bar{y}_i) \equiv -3\bar{y}_i{}^2 + b\bar{y}_i < 0 \qquad (6.40)$$

$\eta = \eta_0 \exp(\lambda t)$. It follows from the plot of the function $Q(y)$ and equation (6.40) that $\bar{y}_1$, $\bar{y}_3$ are stable, whereas $\bar{y}_2$ is an unstable stationary state, $\bar{y}_1 < \bar{y}_2 < \bar{y}_3$. Returning in (6.37) to the x variable we obtain

$$Q(\bar{x}_i) \equiv -\bar{x}_i{}^3 + \bar{x}_i{}^2 - \beta\bar{x}_i + \gamma = 0, \quad i = 1, 2, 3 \qquad (6.41)$$

and it follows from the requirements $\beta > 0$, $\gamma > 0$ that all $\bar{x}_i$ values are positive.

The sensitive state corresponds to zero value of λ. From this condition as well as from equation (6.40) for λ and from (6.37) we obtain the condition for an occurrence of the sensitive state of the cusp catastrophe:

$$\Delta \equiv 27a^2 - 4b^3 \equiv 27(\gamma + 2/27 - 1/3\,\beta)^2 - 4(1/3 - \beta)^3 = 0 \qquad (6.42)$$

This is a known condition for the existence of multiple roots of equation (6.37). When $\Delta < 0$, equation (6.37) has three stationary states of which $\bar{y}_1$, $\bar{y}_3$ are stable; on the other hand, when $\Delta > 0$, there is only on stable stationary state. The system crossing the sensitive state changes its properties: from a bistable state it becomes a monostable state.

6.3.1.8 The Edelstein reaction

A mechanism of isomerization whose overall reaction, $A \rightleftharpoons F$, is of the same form as in the case of the Schloegl reaction,

$$A + X \underset{k_{-1}}{\overset{k_1}{\rightleftharpoons}} 2X \tag{6.43a}$$

$$X + Y \underset{k_{-2}}{\overset{k_2}{\rightleftharpoons}} D \tag{6.43b}$$

$$D \underset{k_{-3}}{\overset{k_3}{\rightleftharpoons}} Y + F \tag{6.43c}$$

is called the Edelstein reaction. The Edelstein mechanism may be regarded as a generalization of the Michaelis–Menten mechanism, which can be represented by the sequence of reactions

$$A + Y \underset{k'_{-a}}{\overset{k'_a}{\rightleftharpoons}} D \tag{6.43a'}$$

$$D \xrightarrow{k_1} Y + F \tag{6.43b'}$$

Equation (6.43a) describes an autocatalytic production of X from A, while (6.43b), (6.43c) represent an enzymatic degradation of X to F proceeding with a regeneration of the enzyme Y. The intermediate compound D is a complex, $D = XY$. Hence, the total enzyme concentration, e, is given by

$$e = [Y] + [D] \equiv y + d \tag{6.44a}$$

Let us assume that during reaction (6.43) the concentrations $[A] \equiv a$, $[F] \equiv f$ and $[Y] + [D]$ are maintained at a constant level:

$$a = \text{const}, \quad f = \text{const}, \quad e = y + d = \text{const} \tag{6.44b}$$

while only $x \equiv [X]$, $y \equiv [Y]$ and $d \equiv [D]$ undergo changes. Changes of independent concentrations x, y (d is related to y via relations (6.44a)) are thus described by the following system of equations:

$$dx/dt = k_1 ax - k_{-1}x^2 - k_2 xy + k_{-2}y \tag{6.45a}$$

$$dy/dt = -k_2 xy - (k_{-2} + k_3)d + k_{-3}fy \tag{6.45b}$$

with conditions (6.44b), where $x = [X]$, $y = [Y]$, $a = [A]$, $d = [D]$, $f = [F]$.
Introducing dimensionless variables by means of the substitution

$$t \to (k_{-1}k_{-2}/k_2)t, \quad x \to (k_2/k_{-2})x, \quad y \to [k_2^2/(k_{-1}k_{-2})]y$$
$$a \to [k_{-1}k_{-2}/(k_1 k_2)]a, \quad f \to (k_{-2}k_{-3})f, \quad e \to (k_{-1}k_{-2}/k_2^2)e$$
$$\varkappa = k_2/k_{-1}, \quad \Lambda = k_3/k_{-2} \tag{6.46}$$

we obtain from the system (6.45):

$$\dot{x} = e - y + ax - x^2 - xy \tag{6.47a}$$

$$\dot{y} = \varkappa\left[(1 + \varLambda)e - (1 + \varLambda + f)y - xy\right] \tag{6.47b}$$

Stationary states of the system (6.47) satisfy the equations resulting from condition (5.4)

$$F(\bar{x}) \equiv \bar{x}^3 + (1 + \varLambda + f - a)\bar{x}^2 + \left[\varLambda e - a(1 + \varLambda + f)\right]\bar{x} - fe \tag{6.48a}$$

$$\bar{y} = (1 + \varLambda)e/\left[\bar{x} + (1 + \varLambda + f)\right] \tag{6.48b}$$

Equation (6.48a) can have one or three real roots, only positive roots having a meaning. We will examine properties of the Edelstein system assuming that control parameters vary in such a region that all roots of (6.48a) are positive.

When investigating the Edelstein reaction we will also assume that $\varkappa \gg 1$ (the case $\varkappa \ll 1$ can be analysed in the same way). This assumption means that the rate constant of forward reaction (6.43b) is much larger than that of reverse reaction (6.43a). With this assumption the system of kinetic equations (6.47) can be written in the following form:

$$\dot{x} = e + ax - y - x^2 - xy \tag{6.49a}$$

$$\varepsilon\dot{y} = (1 + \varLambda)e - (\dot{1} + \varLambda + f)y - xy \tag{6.49b}$$

where $\varepsilon \equiv \varkappa^{-1} \ll 1$. Application of the Tikhonov theorem to (6.49) yields

$$\dot{x} = e + ax - y - x^2 - xy \tag{6.50a}$$

$$0 = (1 + \varLambda)e - (1 + \varLambda + f)y - xy \tag{6.50b}$$

or

$$\dot{x} = G(x) \equiv -F(x)/(1 + \varLambda + f + x) \tag{6.51}$$

where $F(x)$ is defined in (6.48a). As the function $F(x)$ is a third-order polynomial with positive roots, the function $G(x)$ may be replaced by a third-order polynomial by expanding the denominator in a series and preserving only the first term (validity of such an approach follows from structural stability of the cusp catastrophe, see Chapter 2). In other words, the Edelstein kinetic system is equivalent, with the above assumptions, to the Schloegl system.

6.3.1.9 The Lotka model

The first theoretical model of a chemical reaction providing for oscillations in concentrations of reagents was the Lotka model from 1910.
A mechanism of the hypothetical Lotka reaction has the following form:

$$A + X \xrightarrow{k_1} 2X \tag{6.52a}$$

$$X + Y \xrightarrow{k_2} 2Y \tag{6.52b}$$

$$Y \xrightarrow{k_3} P \tag{6.52c}$$

where the concentration of the reactant A is maintained constant, the
product P has no effect on the reaction kinetics and all the reactions are
irreversible. Note that two autocatalytic reactions (6.52a), (6.52b) occur in
the Lotka model. In Chapter 4 we proved that autocatalytic reactions may
appear as slow dynamics in a standard kinetic system.

The Lotka model also has an ecological intepretation. In such a formulation A is an inexhaustible source of food for animals X which, in turn,
constitute feed for animals Y. The death rate of predators Y, denoted as P, is
proportional to their number whereas the death rate of animals X can be
neglected, since a decrease in the abundance of population X occurs mainly
as a result of predators eating them. We must also assume that the
considered populations are isolated from other predators and other animals
being a prey. A model system may consist of hares (X) and lynxes (Y).
Abundant populations of these animals, living in the same area, occur for
example in Canada.

We shall now proceed to giving a system of kinetic equations corresponding
to the mechanism (6.52). Neglecting the effects associated with diffusion, the
kinetic equations take the following form:

$$\dot{x} = k_1 ax - k_2 xy \tag{6.53a}$$

$$\dot{y} = k_2 xy - k_3 y \tag{6.53b}$$

where the designations: $a = [A]$, $x = [X]$, $y = [Y]$ were introduced.

Transforming (6.53) into dimensionless variables

$$t \rightarrow (1/k_3)t, \quad x \rightarrow (k_3/k_2)x, \quad y \rightarrow (ak_1/k_2)y, \quad a \rightarrow (k_3/k_1)a \tag{6.54}$$

we obtain

$$\dot{x} = ax - axy \tag{6.55a}$$

$$\dot{y} = -y + xy \tag{6.55b}$$

Stationary states of the system (6.55) satisfy for $a > 0$ the equations deriving from condition (5.4)

$$\bar{x}(1 - \bar{y}) = 0 \tag{6.56a}$$

$$\bar{y}(\bar{x} - 1) = 0 \tag{6.56b}$$

and always have two solutions

$$\bar{x}_1 = 0, \quad \bar{y}_1 = 0 \tag{6.57a}$$

$$\bar{x}_2 = 1, \quad \bar{y}_2 = 1 \tag{6.57b}$$

Hence, a catastrophe involving the change in a number of stationary solutions cannot occur.

Linearization of (6.55) nearby the stationary point (6.57a) yields (see equations (5.5), (5.6))

$$\dot{\xi} = a\xi \tag{6.58a}$$

$$\dot{\eta} = -\eta \tag{6.58b}$$

the eigenvalues of the stability matrix are thus equal to

$$\lambda_1 = a > 0, \quad \lambda_2 = -1 < 0 \tag{6.59}$$

which corresponds to an unstable stationary point of the saddle type. This stationary state cannot, for every $a > 0$, change type, i.e. a catastrophe cannot take place.

Linearization of (6.55) in the neighbourhood of the stationary point (1, 1) leads to the following system of equations:

$$\dot{\xi} = -a\eta \tag{6.60a}$$

$$\dot{\eta} = +\xi \tag{6.60b}$$

The characteristic equation (5.9)

$$\lambda^2 + a = 0 \tag{6.61}$$

has a pair of imaginary roots, $\lambda_{1,2} = \pm\sqrt{a}\, i$, which corresponds to the stationary state of a centre type. Hence, it is a sensitive state, structurally unstable for all values of the parameter $a > 0$. Thus, the Lotka system (without diffusion) cannot represent a real physical system. On the other hand, Murray has demonstrated the Lotka system to be structurally stable when diffusion is taken into account.

The Lotka equations can be solved exactly. Dividing the two equations we obtain an equation with separated variables which can be integrated, yielding the constant of motion $H(x, y)$

$$dy/dx = y(-1 + x)/ax(1 - y) \tag{6.62}$$

$$H(x, y) = -1/ax - y + 1/a\ln(x) + \ln(y) = \text{const} \tag{6.63}$$

For a set H value the solution (6.63) describes a closed trajectory in the (x, y) phase plane, see Fig. 96.

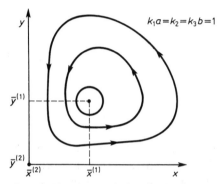

Fig. 96. Phase portrait for the Lotka system.

It is worth adding that in the studied populations of hares and lynxes, occurring in Canada, oscillations in their abundance have been actually observed.

An existence of the constant of motion means that the Lotka system is structurally unstable. Namely, it can be written in a Hamiltonian form and, as we already explained in Chapter 5, the Hamiltonian systems are structurally unstable.

By transforming variables $1/a\ln(x) = u$, $\ln(y) = v$ in (6.55) we indeed obtain a Hamiltonian system with the Hamiltonian $H(u, v)$

$$\dot{u} = 1 - e^v \tag{6.64a}$$

$$\dot{v} = -1 + e^{au} \tag{6.64b}$$

$$H(u, v) = u + v - 1/a\, e^{au} - e^v \tag{6.65}$$

where $H(u, v) = H(x, y)$.

6.3.1.10 The Sel'kov model

The Sel'kov model describes a real chemical system in which oscillations of concentrations of components have been observed. A typical biological system being investigated is the yeast extract. A mechanism of the reaction corresponds to a certain enzymatic reaction, glycolysis, in which a molecule of sugar is cleaved by an enzyme E. In the course of the reaction a low-energy phosphate compound, ADP, is converted into a high-energy compound, ATP.

The Sel'kov mechanism may be written in the following form:

$$ADP + E(ATP)_\gamma \underset{k_{-1}}{\overset{k_1}{\rightleftharpoons}} (ADP)E(ATP)_\gamma \tag{6.66a}$$

$$(ADP)E(ATP)_\gamma \overset{k_2}{\longrightarrow} E(ATP)_\gamma + ATP \tag{6.66b}$$

$$\gamma ATP + E \underset{k_{-3}}{\overset{k_3}{\rightleftharpoons}} E(ATP)_\gamma \tag{6.66c}$$

where the parameter $\gamma > 1$.

The system (6.66) is open: in reaction (6.66a) ADP is fed at a constant rate and ATP released in reaction (6.66b) is removed from the reaction medium. Furthermore, stirring is assumed to be sufficiently effective so that the effects associated with diffusion can be disregarded.

After making a number of assumptions on the magnitude of rate constants of the reaction and on the initial concentrations Sel'kov reduced, using the Tikhonov theorem, the system of kinetic equations corresponding to the mechanism (6.66), bringing it to the following final form:

$$\dot{x} = 1 - xy^\gamma \tag{6.67a}$$

$$\dot{y} = \alpha y(xy^{\gamma-1} - 1) \tag{6.67b}$$

In equations (6.67) $x = [ADP]$, $y = [ATP]$; α and γ are control parameters, $0 < \alpha$, $1 < \gamma$. The Sel'kov model can be regarded as a generalization of the Lotka model, since for $\gamma = 1$ and $x \cong 1$ equations (6.67) are a good approximation to equations (6.55).

The system (6.67) has just one stationary state:

$$\bar{x} = 1, \quad \bar{y} = 1 \tag{6.68}$$

the same as the second stationary state of the Lotka model.

Substituting in (6.67) for $x = \bar{x} + \xi$, $y = \bar{y} + \eta$, we obtain, provided that $|\xi| < 1$, $|\eta| < 1$,

$$\dot{\xi} = -\xi - \gamma\eta - \gamma\xi\eta - \gamma(\gamma - 1)/2\eta^2 + \ldots \tag{6.69a}$$

$$\dot{\eta} = \alpha\xi + \alpha(\gamma - 1)\eta + \alpha\gamma\eta + \alpha\gamma(\gamma - 1)/2\eta^2 + \ldots \tag{6.69b}$$

Neglecting the terms of second and higher orders we obtain linear equations

$$\dot{\xi} = -\xi - \gamma\eta \tag{6.70a}$$

$$\dot{\eta} = \alpha\xi + \alpha(\gamma - 1)\eta \tag{6.70b}$$

The eigenvalues $\lambda_{1,2}$ of the stability matrix satisfy the characteristic equation (5.9)

$$\lambda^2 + [1 - \alpha(\gamma - 1)]\lambda + \alpha = 0 \tag{6.71}$$

$$\lambda_{1,2} = 1/2[\alpha(\gamma - 1) - 1] \pm 1/2\{[\alpha(\gamma - 1) - 1]^2 - 4\alpha\}^{1/2} \tag{6.72}$$

Several sensitive states are possible in the Sel'kov system: two corresponding to the situation $\lambda_1 = \lambda_2$ and one of the $\mathrm{Re}(\lambda_{1,2})$ type.

When $0 < \alpha < \alpha_1 \equiv [(\sqrt{\gamma} - 1)/(\gamma - 1]^2$ the eigenvalues $\lambda_{1,2}$ are real negative and the stationary state $(1, 1)$ is a stable node. When $\alpha = \alpha_1$ equation (6.71) has two equal real roots, $\lambda_1 = \lambda_2 < 0$. The value λ_1 is determined from the requirement of vanishing of the discriminant Δ of the quadratic equation (6.71). This is a sensitive state: when α increasing exceeds the value α_1, a change in the phase portrait of the system (6.67) takes place: the stationary state is, for $\alpha_1 < \alpha < \alpha_0 \equiv (\gamma - 1)^{-1}$, a stable focus (the real part of $\lambda_{1,2}$ is negative). In accordance with what we established in Chapter 5, the catastrophe does not alter a phase portrait nearby a stationary point but is of a global character.

The second sensitive state is that corresponding to $\mathrm{Re}(\lambda_{1,2}) = 0$, that is $\alpha = \alpha_0 = (\gamma - 1)^{-1}$. For $\alpha_0 < \alpha < \alpha_2 \equiv [(\sqrt{\gamma} + 1)/(\gamma - 1)]$ the stationary state is an unstable focus, hence the sensitive state corresponds to the Hopf bifurcation. This catastrophe will be examined in more detail. The state $\alpha = \alpha_2$ is a third sensitive state: for this α value we have $\lambda_1 = \lambda_2 > 0$ (the discriminant of equation (6.71) vanishes) and for $\alpha_2 < \alpha$ the stationary state is an unstable node. This catastrophe has a global character and is analogous to the catastrophe occurring for the value $\alpha = \alpha_1$.

We will now return to a detailed investigation of the possibility of appearance of the Hopf bifurcation for $\alpha = \alpha_0 = (\gamma - 1)^{-1}$. For this purpose equations (6.69) will be written in the form

$$\dot{\xi} = -\xi - \gamma\eta + \{-\gamma\xi\eta - \gamma(\gamma - 1)/2\eta^2 + ...\} \tag{6.73a}$$

$$\dot{\eta} = \alpha\xi + \eta + \{\varepsilon\eta + \alpha\gamma\xi\eta + \alpha\gamma(\gamma - 1)/2\eta^2 + ...\} \tag{6.73b}$$

where $\varepsilon = \alpha(\gamma - 1) - 1$. The expression in braces will be regarded as a perturbation of a Hamiltonian system with the Hamiltonian H

$$\dot{\xi} = -\xi - \gamma\eta = \partial H/\partial\eta \tag{6.74a}$$

$$\dot{\eta} = \alpha\xi + \eta = \partial H/\partial\xi \tag{6.74b}$$

$$H = 1/2(\alpha\xi^2 + 2\xi\eta + \gamma\eta^2) \tag{6.75}$$

which, for $\alpha \cong (\gamma - 1)^{-1}$ and $\gamma > 1$, is positive and represents the energy of the system. The trajectories of the system (6.74), $H = \text{const}$, are for $\alpha \cong (\gamma - 1)$ ellipses

$$\alpha^{1/2}\xi + \alpha^{-1/2}\eta = R\cos(t) \tag{6.76a}$$

$$[(\alpha\gamma - 1)/\alpha]^{1/2}\eta = R\sin(t) \tag{6.76b}$$

We will now compute the derivative of H with respect to time on the trajectories of the system (6.73), according to equations given in Appendix A4.

$$\mathring{H} = +\varepsilon(\xi + \gamma\eta)\eta - \gamma(1 - \alpha\gamma)\{\xi + [(\gamma - 1)/2]\eta\}\eta^2 + ... \tag{6.77}$$

In turn, the change in energy, ΔH, on the approximate trajectories (6.76) is computed, yielding

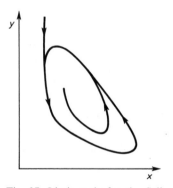

Fig. 97. Limit cycle for the Selkov system.

$$\Delta H = \int_0^{2\Pi} \overset{\circ}{H} \, dt \cong \varepsilon(\gamma - 1)\left\{\pi R^2 - 1/4[\alpha/\alpha\gamma - 1]\right\}^3 R^4\}$$ (6.78)

For $\varepsilon < 0$, $\varepsilon \equiv \alpha(\gamma - 1) - 1$, and for small R, $\Delta H < 0$ and the stationary state (6.68) is stable. In contrast, for $\varepsilon > 0$ and for small R, $\Delta H > 0$. This implies that the stationary state (6.68) loses stability for $\alpha_c > (\gamma - 1)^{-1}$. For $R > R_{cr} = [4\pi(\gamma - \alpha^{-1})^3]^{1/2}$, $\Delta H < 0$, that is the limit cycle is stable.

Equations (6.76) describe an approximate form of the limit cycle, $R = R_{cr}$, see Fig. 97.

6.3.1.11 The Brusselator

Another hypothetical mechanism of a chemical reaction is the model called Brusselator, investigated by the Brussels school of Prigogine

$$A \xrightarrow{k_1} X$$ (6.79a)

$$B + X \xrightarrow{k_2} Y + C$$ (6.79b)

$$2X + Y \xrightarrow{k_3} 3X$$ (6.79c)

$$X \xrightarrow{k_4} D$$ (6.79d)

where concentrations of the substances A, B, C are maintained constant and the product D has no effect on the reaction kinetics. Equations (6.79) are assumed to be irreversible. In the model occurs a termolecular reaction (6.79c), which is additionally of an autocatalytic nature. However, we demonstrated in Chapter 4 that such reactions may appear as slow dynamics of a standard kinetic system. Some other mechanisms, discussed in a book by Edelstein and leading to the same kinetic equations, are related to the mechanism (6.79).

The following system of kinetic equations without diffusion corresponds to the mechanism (6.79):

$$\dot{x} = k_1 a + k_3 x^2 y - k_2 bx - k_4 x$$ (6.80a)

$$\dot{y} = -k_3 x^2 y + k_2 bx$$ (6.80b)

where $x = [X]$, $y = [Y]$, $a = [A]$, $b = [B]$.

We shall introduce dimensionless variables, making substitutions in (6.80)

$$t \to t k_4^{-1}, \quad x \to x(k_4/k_2)^{1/2}, \quad y \to y(k_4/k_2)^{1/2}$$
$$a \to a(k_2 k_4)^{1/2} k_1^{-1}, \quad b \to b(k_4/k_3) \tag{6.81}$$

which leads to the system of equations

$$\dot{x} = a - (b+1)x + x^2 y \tag{6.82a}$$

$$\dot{y} = bx - x^2 y \tag{6.82b}$$

An analysis of the system (6.82) will begin with the determination of its stationary states. The requirement $\dot{\bar{x}} = 0$, $\dot{\bar{y}} = 0$ results in the determination of the only stationary state of the system

$$\bar{x} = a, \quad \bar{y} = b/a \tag{6.83}$$

Substituting in (6.82) for $x = \bar{x} + \xi$, $y = \bar{y} + \eta$, we obtain

$$\dot{\xi} = (b-1)\xi + a^2 \eta + (b/a\xi^2 + 2a\xi\eta + \xi^2 \eta) \tag{6.84a}$$

$$\dot{\eta} = -b\xi - a^2 \eta - (b/a\xi^2 + 2a\xi\eta + \xi^2 \eta) \tag{6.84b}$$

and, assuming smallness of ξ, η, $|\xi|$, $|\eta| \ll 1$, the system (6.84) reduces to a linear system:

$$\dot{\xi} = (b-1)\xi + a^2 \eta \tag{6.85a}$$

$$\dot{\eta} = -b\xi - a^2 \eta \tag{6.85b}$$

The eigenvalues are equal to (see equation (5.9))

$$\lambda_{1,2} = 1/2(a^2 - b + 1) \pm 1/2[(a^2 - b + 1)^2 - 4a^2]^{1/2}$$
$$= 1/2(a^2 - b + 1) \pm 1/2\{[(a-1)^2 - b][(a+1)^2 - b]\}^{1/2} \tag{6.86}$$

Sensitive states correspond to the condition $\mathrm{Re}(\lambda_{1,2}) = 0$. The analysis of equation (6.86) leads to a conclusion that for $a > 0$, $b > 0$ the only sensitive state is given by

$$a^2 - b + 1 = 0 \tag{6.87}$$

$$\lambda_{1,2} = \pm a\sqrt{-1} \tag{6.88}$$

When $a^2 - b + 1 < 0$ and $(a-1)^2 < b < (a+1)^2$, the eigenvalues (6.86) have a negative real part and the stationary state corresponds to a stable focus. The case $a^2 - b + 1 = 0$ corresponds to a sensitive state typical for the Hopf bifurcation; when $a^2 - b + 1 > 0$ and $(a-1)^2 < b < (a+1)^2$, the stationary state is an unstable focus.

On the basis of the above analysis we arrive at a conclusion that in the system represented by the equations not accounting for diffusion (6.82) the Hopf bifurcation may appear. We shall examine the catastrophe taking place in more detail using the method described in Section 5.6.

The system (6.84) will be written in the form

$$\dot{\xi} = (b-1)\xi + a^2\eta + \{b/a\xi^2 + 2a\xi\eta + \xi^2\eta\} \tag{6.89a}$$

$$\dot{\eta} = -b\xi - (b-1)\eta - \{-\varepsilon\eta + b/a\xi^2 + 2a\xi\eta + \xi^2\eta\} \tag{6.89b}$$

where the following substitution was introduced into (6.89)

$$\varepsilon \equiv b - a^2 - 1 \tag{6.90}$$

We will assume that $|\varepsilon| \ll 1$, since the system is examined nearby the sensitive state (6.87). We will also assume that $|\varepsilon| < a^2$, i.e. $b > 1$.

As an unperturbed system we will take the canonical system

$$\dot{\xi} = (b-1)\xi + a^2\eta \tag{6.91a}$$

$$\dot{\eta} = -b\xi - (b-1)\eta \tag{6.91b}$$

with the Hamiltonian

$$H(\xi, \eta) = 1/2[b\xi^2 + 2(b-1)\xi\eta + a^2\eta^2] > 0 \tag{6.92}$$

Hence, the phase trajectories of the system (6.91), $H = \text{const}$, are ellipses

$$[(b-1)/a]\xi + a\eta = R\cos(t) \tag{6.93a}$$

$$\{b - [(b-1)/a]^2\}\xi = R\sin(t) \tag{6.93b}$$

Let us now compute the derivative of the Hamiltonian (6.92) on the trajectories of the system (6.89):

$$\overset{\circ}{H} = [b\xi + (b-1)\eta]\dot{\xi} + [(b-1)\xi + a^2\eta]\dot{\eta} \tag{6.94}$$

$$\Delta H = \int_0^{2\Pi} \overset{\circ}{H} dt \cong \varepsilon(b-1)\pi R^2 - 3/4\,\pi R^4 \tag{6.95}$$

When $1 < b < a^2 + 1$, the parameter $\varepsilon < 0$ and the stationary state $(a, b/a)$ is stable. For $b > a^2 + 1$ the stationary state is no longer stable but a stable limit cycle is generated: $\Delta H < 0$ for $R > R_{cr} \cong$ $\cong [4/3(b-1)(b-a^2-1)]^{1/2}$. An approximate shape of the limit cycle is given by equation (6.93), $R = R_{cr}$, see Fig. 98.

To end with, note also that the stability of the limit cycle results from the

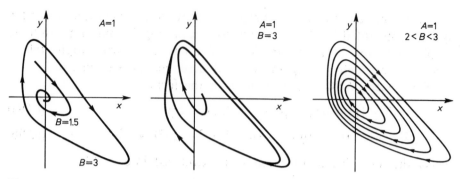

Fig. 98. Limit cycle in the Brusselator.

presence in (6.95) of the term proportional to R^4 which originates from the third-order terms in equations (6.89) (or (6.82)). The neglect of these terms would lead to a false conclusion about the nonexistence of a limit cycle in the Brusselator.

6.3.1.12 The Oregonator

A simplified mechanism of the Belousov–Zhabotinskii reaction, proposed by Fields and Noyes (Oregonator), was described in Section 6.2.4, see equations (6.13).

We assume that [A] is maintained at a constant level, the reactions are irreversible and P has no effect on the reaction kinetics.

Kinetic equations corresponding to the mechanism (6.13) are of the form

$$\dot{x} = k_1 a y - k_2 x y + k_3 a x - 2 k_4 x^2 \tag{6.96a}$$

$$\dot{y} = -k_1 a y - k_2 x y + f k_5 z \tag{6.96b}$$

$$\dot{z} = 2 k_3 a x - k_5 z \tag{6.96c}$$

Transforming in (6.96) into dimensionless variables:

$$t \to (1/k_1 a) t, \quad \varepsilon \to k_1/k_3, \quad p = k_1 a/k_5, \quad q = 2 k_1 k_4/(k_2 k_3)$$

$$x \to (k_1 a/k_2) x, \quad y \to (k_3 a/k_2) y, \quad z \to (2 k_1 k_3 a^2/k_2 k_5) z \tag{6.97}$$

we obtain

$$\varepsilon \dot{x} = x + y + -xy - q x^2 \tag{6.98a}$$

$$\dot{y} = 2 f z - y - x y \tag{6.98b}$$

$$p \dot{z} = x - z \tag{6.98c}$$

On the basis of known values of the constants $k_1, ..., k_5$, see Table 6.5, Section 6.2.4, values of the constants ε, p, q can be estimated:

$$\varepsilon \cong 2 \times 10^{-4}, \quad p \cong 3.1 \times 10^2, \quad q \cong 8.4 \times 10^{-6} \tag{6.99}$$

It is thus seen that in the model (6.98) time scales are clearly marked.

In the case of carrying out the reaction in an open reactor, a constant stream of reagents entering the reactor should be taken into account in (6.96) (see Chapter 4):

$$\dot{x} = k_1 a y - k_2 x y + k_3 a x - 2k_4 x^2 + 1/T(x_0 - x) \tag{6.96a*}$$

$$\dot{y} = -k_1 a y - k_2 x y + f k_5 z + 1/T(y_0 - y) \tag{6.96b*}$$

$$\dot{z} = 2k_3 a x - k_5 z + 1/T(z_0 - z) \tag{6.96c*}$$

where T is the retention time.

Stationary solutions to the system (6.98), $\dot{x} = 0$, $\dot{y} = 0$, $\dot{z} = 0$, having a chemical meaning, $\bar{x} \geq 0$, $\bar{y} \geq 0$, $\bar{z} \geq 0$, are given by

$$\bar{x}_1 = 0, \quad \bar{y}_1 = 0, \quad \bar{z}_1 = 0 \tag{6.100a}$$

$$\bar{x}_2 = \{(1 - 2f + q) + [(1 - 2f - q)^2 + 4q(1 + 2f)]^{1/2}\}/(2q)$$

$$\bar{y}_2 = 1/2(1 + 2f - q\bar{x}_2), \quad \bar{z}_2 = \bar{x}_2 \tag{6.100b}$$

where $q \geq 0$, $f \geq 0$.

We shall examine a stability of the stationary state $(\bar{x}_1, \bar{y}_1, \bar{z}_1)$ by substituting in (6.98) for $x = \xi$, $y = \eta$, $z = \zeta$, $|\xi| \ll 1$, $|\eta| \ll 1$, $|\zeta| \ll 1$, which yields

$$\varepsilon\dot{\xi} = \xi + \eta \tag{6.101a}$$

$$\dot{\eta} = -\eta + 2f\xi \tag{6.101b}$$

$$p\dot{\zeta} = \xi - \zeta \tag{6.101c}$$

The eigenvalues of the stability matrix satisfy the characteristic equation (5.9)

$$\varepsilon p \lambda^3 - (p - \varepsilon - \varepsilon p)\lambda^2 - (p + 1 - \varepsilon) - (1 + 2f) = 0 \tag{6.102}$$

The product of the roots of equation (6.102), $(1 + 2f)/(\varepsilon p)$, is positive. Consequently, there exists at least one root with a positive real part. Accordingly, the stationary state $(\bar{x}_1, \bar{y}_1, \bar{z}_1) = (0, 0, 0)$ is unstable. We will not examine the possibility of appearance of a limit cycle from this stationary state by way of the Hopf bifurcation, since the limit cycle would have to partially exist in the region of negative concentrations.

We shall now investigate a stability of the stationary state $(\bar{x}_2, \bar{y}_2, \bar{z}_2)$, substituting in (6.98) for $x = \bar{x}_2 + \xi$, $y = \bar{y}_2 + \eta$, $z = \bar{z}_2 + \zeta$, and obtaining:

$$\varepsilon\dot{\xi} = -a\xi - b\eta - q\xi^2 - \xi\eta \tag{6.103a}$$

$$\dot{\eta} = -c\xi - d\eta + 2f\zeta - \xi\eta \tag{6.103b}$$

$$p\dot{\zeta} = \xi - \zeta \tag{6.103c}$$

$$a = -1 + 2q\bar{x}_2 + \bar{y}_2 = q\bar{x}_2 + \bar{y}_2/\bar{x}_2$$

$$b = \bar{x}_2 - 1, \quad c = \bar{y}_2, \quad d = \bar{x}_2 + 1 \tag{6.104}$$

The parameters c, d are always positive, $b > 0$ if $q < 1$ whereas a is always positive which follows from (6.100b). If we now assume that $|\xi| \ll 1$, $|\eta| \ll 1$, $|\zeta| \ll 1$, the linear system

$$\varepsilon\dot{\xi} = -a\xi - b\eta \tag{6.105a}$$

$$\dot{\eta} = -c\xi - d\eta + 2f\zeta \tag{6.105b}$$

$$p\dot{\zeta} = \xi - \zeta \tag{6.105c}$$

is obtained.

The eigenvalues of the stability matrix fulfil the characteristic equation (5.9)

$$\lambda^3 + A\lambda^2 + B\lambda + C = 0 \tag{6.106a}$$

$$A = D + 1/p, \quad B = D/p + [2q\bar{x}_2 + \bar{x}_2(q - 1) + 2f]/e$$

$$C = \bar{x}_2(2q\bar{x}_2 + q + 2f - 1)/(\varepsilon p)$$

$$D = q\bar{x}_2/\varepsilon + \bar{y}_2/(\varepsilon\bar{x}_2) + \bar{x}_2 + 1 > 0 \tag{6.106b}$$

The stationary state $(\bar{x}_2, \bar{y}_2, \bar{z}_2)$ will be stable when all the roots of equation (6.106) have negative real parts. We will investigate the conditions under which this stationary state loses stability, that is under which at least one solution with a positive real part appears. Next, in the region of control parameters corresponding to instability of the state $(\bar{x}_2, \bar{y}_2, \bar{z}_2)$ we shall examine possible catastrophes of codimension $\leqslant 2$. It follows from the classification given in Section 5.5 that the bifurcations of codimension one and two of a sensitive state corresponding to the requirement $\lambda_1 = 0$ are theoretically possible; the Hopf bifurcation for which a sensitive state is of the form $\lambda_1 = \alpha$, $\lambda_{2,3} = \pm i\beta$ and the Takens–Bogdanov bifurcation with a sensitive state having the corresponding eigenvalues of the stability matrix $\lambda_1 = 0$, $\lambda_{2,3} = \pm i\beta$ are also possible.

First, let us examine the possibility of the appearance of the bifurcations of codimension one and two associated with the sensitive state $\lambda_1 = 0$. Such a sensitive state is represented by equation (6.106a) in which the coefficient C, proportional to the product $\lambda_1 \lambda_2 \lambda_3$, is equal to zero. Since the parameter C, owing to inequality (6.106) cannot be zero, $C > 0$, catastrophes of codimension one and two, having the sensitive state $\lambda_1 = 0$, can be excluded.

We shall now examine the case of the Bogdanov bifurcation. The sensitive state $\lambda_1 = 0$, $\lambda_{2,3} = \pm i\beta$ is represented by the eigenequation

$$\lambda^3 + \beta^2 \lambda = 0 \tag{6.107}$$

Comparison of the coefficients of equations (6.107), (6.106) yields the requirements for the sensitive state for the Bogdanov bifurcation: $A = 0$, $B > 0$, $C = 0$.

By virtue of inequality (6.106b) the parameter C (and A) cannot take the value $C = 0$ ($A = 0$), $C > 0$ ($A > 0$).

Recapitulating, in the model under study the stationary state $(\bar{x}_2, \bar{y}_2, \bar{z}_2)$ cannot lose the stability by way of the Bogdanov bifurcation.

We shall now investigate the case of the Hopf bifurcation. A sensitive state is given by the conditions $\lambda_1 = \alpha < 0$, $\lambda_{2,3} = \pm i\beta$ and represented by the eigenequation

$$\lambda^3 - \alpha \lambda^2 + \beta^2 \lambda - \alpha \beta^2 = 0 \tag{6.108}$$

Comparison of the coefficients of equations (6.108), (6.106) yields the conditions for a sensitive state of the Hopf bifurcation

$$A > 0, \quad B > 0, \quad C > 0, \quad AB = CD \tag{6.109}$$

The necessary and sufficient conditions for the solutions λ_1, λ_2, λ_3 to equation (6.106) to have negative real parts, which corresponds to a stability of the stationary state, result from the Hurwitz criterion (see Appendix A5.8)

$$A > 0, \quad C > 0, \quad AB - C > 0 \tag{6.110}$$

where, because $A > 0$, $C > 0$, the last inequality requires meeting the condition $B > 0$. If any of the conditions (6.110) is not satisfied, the stationary state $(\bar{x}_2, \bar{y}_2, \bar{z}_2)$ loses stability.

As can be seen, the fact of the stationary state losing stability does not necessarily imply the appearance of conditions (necessary) for a limit cycle

(Hopf bifurcation) to occur. The stability of a stationary point must be perturbed in such a way that on changing control parameters the last inequality in (6.110) becomes an equality and then again an inequality with a sense reversed.

The conditions $A > 0$, $C > 0$ are always fulfilled by virtue of the definition (6.102b). Hence, critical values of the parameters follow from the equality $AB = C$.

Substitution of definitions of A, B, C into the equality $AB \geqslant C$ leads to the requirement

$$0 < 1/p \leqslant -1/2D\left[D^2 + 2f/\varepsilon(1 - \bar{x}_2)\right] +$$
$$+ 1/2D\left\{\left[D^2 + 2f/\varepsilon(1 - \bar{x}_2)\right]^2 - \right.$$
$$\left. - (4D^2/\varepsilon)\left[2q\bar{x}_2{}^2 + (q - 1)\bar{x}_2 + 2f\right]\right\}^{1/2} \tag{6.111}$$

with the critical case corresponding to the equality:

$$1/p = -1/2D\left[D^2 + 2f/\varepsilon(1 - \bar{x}_2)\right] + 1/2D\left\{\left[D^2 + 2f/\varepsilon(1 - \bar{x}_2)\right]^2 - \right.$$
$$\left. - (4D^2/\varepsilon)\left[2q\bar{x}_2{}^2 + (q - 1)\bar{x}_2 + 2f\right]\right\}^{1/2} \tag{6.111'}$$

Substituting into equality (6.111') approximate values of the parameters p, ε, q, see equalities (6.99), we may compute critical values of the unknown parameter f (there can be several solutions to (6.111') satisfying the condition $f > 0$).

Murray has shown that the critical values of the parameter f are bounded by the values f_1, f_2, approximately equal to

$$f_1 \cong 0.25, \quad f_2 \cong 1.206 \tag{6.112}$$

and he determined from inequality (6.111) the region in the $(f, 1/p)$ plane within which the sensitive state (states) of the Hopf bifurcation exist, see Fig. 99.

From the above analysis a conclusion can be drawn that, because the stationary state $(\bar{x}_2, \bar{y}_2, \bar{z}_2)$ can have a sensitive state corresponding at most to the Hopf bifurcation, the centre manifold theorem may be applied to the system of equations (6.98) thus reducing it to a system of equations in two variables in which the Hopf bifurcation would appear.

The system (6.98) has a form well suited for an application of the Tikhonov theorem and we already know that by reducing the system to a system in two variables we do not lose any catastrophe connected with the loss of stability of the stationary point $(\bar{x}_2, \bar{y}_2, \bar{z}_2)$. It follows from equality

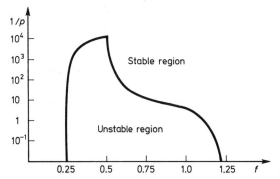

Fig. 99. Sensitive states of the Hopf bifurcation in the Oregonator.

(6.99) that equation (6.98a) describes a fast process. Hence, from the smallness of the parameter ε follows the approximate equation

$$x + y - xy - qx^2 = 0 \tag{6.113}$$

which can be used to eliminate from (6.98b), (6.98c) the x-variable. Using the system of two equations obtained in this way and the method described in Section 5.6 it can be proved that the Oregonator indeed has a limit cycle in the yz-plane: the concentrations y, z vary periodically with time.

6.3.1.13 The Bykov model

One of the mechanisms considered by Bykov is the following sequence of reactions:

$$Z \underset{k_{-1}}{\overset{k_1}{\rightleftharpoons}} X \tag{6.114a}$$

$$X + 2Z \xrightarrow{k_2} 3Z \tag{6.114b}$$

$$Z \underset{k_{-3}}{\overset{k_3}{\rightleftharpoons}} Y \tag{6.114c}$$

The mechanism (6.114) is represented by the following kinetic equations:

$$\dot{x} = k_1 z - k_{-1} x - k_2 x z^2 \tag{6.115a}$$

$$\dot{y} = k_3 z - k_{-3} y \tag{6.115b}$$

$$z = 1 - x - y \tag{6.115c}$$

where $x = [X]$, $y = [Y]$, $z = [Z]$. Equation (6.115c) signifies that the total concentration $x + y + z$ is constant and equal to one.

We shall begin with the substantiation of a certain general property of the system (6.115). It follows from equations (6.115) that for small positive x, y the inequalities $\dot{x} > 0$, $\dot{y} > 0$ are satisfied, i.e. x, y increase. On the other hand, for $(x + y) \gg 1$ it follows from (6.115) that $\dot{x} < 0$, $\dot{y} < 0$ and hence x, y decrease. More detailed considerations reveal that phase trajectories of the system (6.115) remain, for $t \to \infty$, within the limit set

$$D_2\{(x, y): \quad 0 \leqslant x, x + y \leqslant 1\} \tag{6.116}$$

Note that when $k_3 \cong 0$, $k_{-3} \cong 0$, the y-variable may be regarded as a parameter, $y \cong A$, $A = $ const. The system (6.115) is then reduced, by virtue of the Tikhonov theorem, to the equation

$$\dot{x} = k_1[(1 - A) - x] - k_{-1}x - k_2x[(1 - A) - x]^2 = 0 \tag{6.117}$$

which can be brought, using the substitution $x \to x + B$, $B = 2/3\,(1 - A)$, to a standard form of the cusp catastrophe, See section 5.5.3.2. The system (6.115) may thus represent the bistability phenomenon, in analogy with the Schloegl and Edelstein models.

We shall now examine the possibility of occurrence of oscillations in the system (6.115). Stationary states of the system (6.115) satisfy the equations:

$$k_1(1 - \bar{x} - \bar{y}) - k_{-1}\bar{x} - k_2\bar{x}(1 - \bar{x} - \bar{y})^2 = 0 \tag{6.118a}$$

$$k_3(1 - \bar{x} - \bar{y}) - k_{-3}\bar{y} = 0 \tag{6.118b}$$

An approximate solution to (6.118) is of the form

$$\bar{x} \cong a/(a + b + c), \quad \bar{y} = b/(a + b + c) \tag{6.119a}$$

$$a = k_1k_{-3}, \quad b = k_{-1}k_3, \quad c = k_{-1}k_{-3} \tag{6.119b}$$

By linearizing the system of equations (6.115) we obtain the characteristic equation,

$$\lambda^2 + A\lambda + B = 0 \tag{6.120a}$$

$$A = k_1 + k_{-1} + k_3 + k_{-3} + k_2c(2a - c)/(a + b + c)^2 \tag{6.120b}$$

$$B = a + b + c + k_2c[(k_3 + k_{-3})c - 2ak_{-3}]/(a + b + c)^2 \tag{6.120c}$$

which can have the sensitive state of the Hopf bifurcation. Indeed, the constants k_i can be selected so as to satisfy the requirements $A = 0$, $B > 0$

for the sensitive state of the Hopf bifurcation. For instance, for $k_1 \cong 0$, $k_3 \cong 0$ we obtain the approximate equation

$$\lambda^2 + \left(k_{-1} + k_{-3} - k_2\right)\lambda + \left(k_{-1}k_{-3} + k_2 k_{-3}\right) = 0 \qquad (6.121a)$$

$$k_1 \cong 0, \quad k_3 \cong 0 \qquad (6.121b)$$

Hence, to the sensitive state of the Hopf bifurcation corresponds the relation between the constants k_{-1}, k_2, k_{-3}

$$k_2 = k_{-1} + k_{-3} \qquad (6.121c)$$

for then it follows from (6.121a) that

$$\lambda^2 = -\left[k_{-1}k_{-3} + k_{-3}\left(k_{-1} + k_{-3}\right)\right] < 0 \qquad (6.122)$$

When $k_{-1} + k_{-3} - k_2 > 0$, the stationary state is a stable focus, whereas for $k_{-1} + k_{-3} - k_2 < 0$ the stationary state is an unstable focus and the Hopf bifurcation may appear.

The trajectories cannot escape to infinity by virtue of the property (6.116); hence, there must be a limit cycle within the region D_2 and the Hopf bifurcation actually takes place, see Fig. 100.

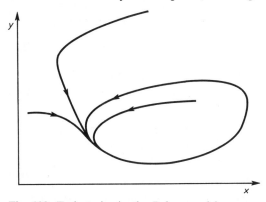

Fig. 100. Trajectories in the Bykov model.

Note in the end that the Takens–Bogdanov bifurcation may occur in the Bykov model, see Section 5.5.3.4. The sensitive state of the Takens–Bogdanov bifurcation, $\lambda_1 = \lambda_2 = 0$, may be obtained by imposing in (6.121a), (6.121b) the following requirements on the constants k_{-1}, k_2, k_{-3}

$$k_2 = k_{-1} \qquad (6.121c')$$

$$k_{-3} = 0 \qquad (6.121d)$$

6.3.2 *Models of reactions with diffusion and their analysis*

6.3.2.1 Introduction

The models of reactions with diffusion will be described below. The Brusselator with diffusion is a model system while the Oregonator with diffusion corresponds to the Belousov–Zhabotinskii reaction. First we shall describe the Fisher–Kolmogorov model with diffusion which, as will be shown later, may be regarded as a special case of the Oregonator with diffusion.

Catastrophes occurring in these models in the presence of diffusion will be examined next. We shall investigate wave phenomena in the Fisher–Kolmogorov model and in the Oregonator as well as dissipative structures in the Brusselator. The studies of systems of reactions with diffusion, both experimental and theoretical, have not led yet to the formulation of a complete theory. Consequently, only fundamental results concerning the phenomena of a loss of stability type in these models will be presented.

6.3.2.2 The Fisher–Kolmogorov model of reactions with diffusion

An equation of a reaction with diffusion for one variable is of the form

$$\partial x/\partial t = P(x) + D\partial^2 x/\partial r^2 \tag{6.123a}$$

where $x = x(t, r)$ and D is a constant diffusion coefficient.

In the case when $P(x) = ax(1 - x)$, $a = $ const, we obtain a model used by Fisher and Kolmogorov for the description of some biological processes (more specifically, the model may represent spread of an epidemic in a certain region or propagation of a favourable gene in a population):

$$\partial x/\partial t = ax(1 - x) + D\partial^2 x/\partial r^2 \tag{6.123b}$$

Note that equation (6.123b) is also obtained by accounting for diffusion in the kinetic equation (6.28) for a reversible autocatalytic reaction and renormalizing the x-variable, $x \to (k_1/k_{-1})x$.

This model also appears in a description of the Belousov–Zhabotinskii reaction as a special case of the Oregonator with diffusion, see Section 6.3.2.3.

Wave processes described by the Fisher–Kolmogorov equation will be considered first. We shall analyse solutions to this equation of the travelling wave form

$$x(t, r) = z_1(r + vt) \equiv z_1(\rho) \tag{6.124}$$

where $v = $ const is the wave velocity. Substitution of (6.141) into (6.123) yields

$$vz_1'(\rho) = az_1(\rho)[1 - z_1(\rho)] + Dz_1''(\rho) \tag{6.125}$$

where prime stands for differentiation with respect to the ρ variable. Introducing a new variable, $z_2 = z_1'$, we obtain from (6.125):

$$z_1' = z_2 \tag{6.126a}$$

$$Dz_2' = -az_1 + vz_2 + az_1^2 \tag{6.126b}$$

Only those solutions have a physical meaning for which $z_1 \geqslant 0$ whereas the sign of the z_2 variable can be arbitrary. Note that the system (6.126) resembles a standard form of the Takens–Bogdanov bifurcation (Section 5.5.3.5), in which $c_1 = -a$, $c_2 = v$, $R_1 = a$, $R_2 = 0$. The difference lies in the fact that in the Takens–Bogdanov bifurcation the parameters R_1, R_2 are fixed and only c_1, c_2 undergo changes. In the case of system (6.126) the parameter $R_1 = -c_1$ is a control parameter.

The system (6.126) has two stationary states, $\bar{z}_1' = 0$, $\bar{z}_2' = 0$,

$$\bar{z}_1^{(1)} = 0, \quad \bar{z}_2^{(1)} = 0 \tag{6.127a}$$

$$\bar{z}_1^{(2)} = 1, \quad \bar{z}_2^{(2)} = 0 \tag{6.127b}$$

Linearization of the system (6.127) nearby the state $(z_1^{(1)}, z_2^{(1)})$,

$$z_1 = \bar{z}_1^{(1)} + \zeta_1 \tag{6.128a}$$

$$z_2 = \bar{z}_2^{(1)} + \zeta_2 \tag{6.128b}$$

leads to a linear system

$$\zeta_1' = \zeta_2 \tag{6.129a}$$

$$D\zeta_2' = -a\zeta_1 + v\zeta_2 \tag{6.129b}$$

The characteristic equation (5.9) corresponding to (6.129) has the roots

$$\lambda_{1,2}^{(1)} = [v \pm (v^2 - 4aD)^{1/2}]/(2D) \tag{6.130}$$

Hence, when $v = 0$ the point $(0, 0)$ is a centre, whereas when $0 < v < 2\sqrt{aD}$ the point $(0, 0)$ is an unstable focus and when $2\sqrt{aD} < v$, the point is an unstable node. The system nearby $(0, 0)$ has two sensitive states: (1) $v = 0 \leftrightarrow \lambda_{1,2}^{(1)} = \pm ia$: (2) $v = 2\sqrt{aD} \leftrightarrow \lambda_1^{(1)} = \lambda_2^{(1)} = \sqrt{a/D}$.

Linearization of (6.126) in the vicinity of the state $(\bar{z}_1^{(2)}, \bar{z}_2^{(2)})$

$$z_1 = \bar{z}_1^{(2)} + \zeta_1 \tag{6.131a}$$

$$z_2 = \bar{z}_2^{(2)} + \zeta_2 \tag{6.131b}$$

leads to a linear system

$$\zeta_1' = \zeta_2 \tag{6.132a}$$

$$D\zeta_2' = a\zeta_1 + v\zeta_2 \tag{6.132b}$$

The characteristic equation corresponding to the system (6.132) has the roots

$$\lambda_{1,2}^{(2)} = \left[v \pm (v^2 + 4aD)^{1/2} \right]/(2D) \tag{6.133}$$

It follows from equation (6.133) that the state $(1, 0)$ is of a saddle type for all v. The system has, nearby the point $(1, 0)$, the third sensitive state: (3) $a = 0 \leftrightarrow \lambda_1^{(2)} = v/D$, $\lambda_2^{(2)} = 0$.

Note that the second sensitive state is of most interest: the requirement $v = 0$ implies the lack of diffusion, while the condition $a = 0$ signifies the lack of a chemical reaction. The sensitive states 1, 3 correspond to catastrophes of the change in a phase portrait in the neighbourhood of a given singular point while the sensitive state 2 represents the catastrophe of a global change of a phase portrait.

The trajectories on a phase plane for the cases (1) $v = 0$, (2) $0 < v < 2\sqrt{aD}$, (3) $2\sqrt{aD} < v$ are shown in Fig. 101.

Let us examine the nature of a catastrophe of a global change in a phase portrait occurring on the crossing by the parameter v of the sensitive state 2, $v = v_{cr} = 2\sqrt{aD}$. When $v < v_{cr}$, the z_1 variable reaches negative values on trajectories — such states from the vicinity of the point $(0, 0)$ have no physical meaning. When $v_{cr} < v$ the states with the initial value $y_0 > 0$ evolve remaining in the positive phase half-plane $z_1 z_2$. Therefore, when $v_{cr} < v$, a wave may appear. As a result, only the waves with a velocity $v \geqslant v_{cr}$ propagate in the system; chemical waves with velocities $v < v_{cr}$ are unstable and vanish.

We will now proceed to the investigation of properties of general solutions, not necessarily of a travelling wave form, to the Fisher–Kolmogorov equation, (6.123b). To simplify further equations, we will perform a change of variables in (6.123b)

$$t \to t/a, \quad r \to rL, \quad D \to DL^2a \tag{6.134}$$

(a)

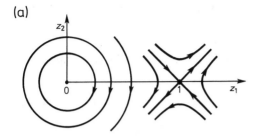

(b)

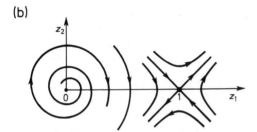

(c)

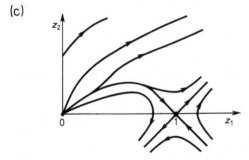

Fig. 101. Phase portrait for the Fisher–Kolmogorov model: (a) $v = 0$; (b) $0 < v < 2\sqrt{aD}$; (c) $2\sqrt{aD} < v$.

where L is the length of a reactor, giving

$$\partial x/\partial t = x(1 - x) + D\partial^2 x/\partial r^2 \qquad (6.135)$$

In accordance with a general procedure described in Chapter 5, we shall seek solutions of (6.135) in the form of a Fourier series

$$x(t,\ r) = \sum_{n}^{\infty} \xi_n(t)\sin(n\pi r) \qquad (6.136)$$

where $|\zeta_n| \ll 1$ and the reactor length varies from $r = 0$ to $r = 1$. The approximation $|\zeta_n| \ll 1$ is good for large D, as then owing to diffusion the concentration x decreases in a given region of the reactor.

Substitution of (6.136) into (6.135) yields linear ordinary differential equations:

$$\dot{\xi}_n = \left(1 - Dn^2\pi^2\right)\xi_n, \quad n = 1, 2, \ldots \tag{6.137}$$

The eigenvalues λ_n, $\xi_n = \xi_{0,n}\,\exp(\lambda_n t)$ are thus equal to

$$\lambda_n = 1 - Dn^2\pi^2 \tag{6.138}$$

The sensitive states corresponding to vanishing of the eigenvalues appear when the diffusion coefficient reaches the critical values

$$D_{\mathrm{cr},n} = 1/(n^2\pi^2) \tag{6.139a}$$

or, making inverse transformation with respect to (6.134),

$$D_{\mathrm{cr},n} = L^2 a/(n^2\pi^2) \tag{6.139b}$$

In other words, when $D > D_{\mathrm{cr},1}$, a loss of stability of the solution $\xi_1(t)$ $\sin(\pi r)$ takes place. The solution increases unlimitedly for $t \to \infty$, whereas all the other solutions $\xi_n(t)\,\sin(n\pi r)$, $n > 1$, approach zero if $D < D_{\mathrm{cr},2}$.

Apparently, to determine an evolution of the solution $\xi_1(t)\,\sin(\pi r)$, a non-linearized, exact equation (6.135) should be examined. However, even from the above analysis follows a conclusion that dissipative structures may appear when $D > D_{\mathrm{cr},1}$.

Note, however, that in the case of a lack of diffusion the stationary state was unstable, see equation (6.138) for $D = 0$. In this case the diffusion stabilized, for $D < D_{\mathrm{cr},1}$, this state. Of considerably more interest is the case of a loss of stability on accounting for diffusion by the stationary state stable in the absence of diffusion. Such an effect of diffusion can be readily shown to be possible only in the case of a system of two equations with diffusion. A catastrophe of this type, called the Turing bifurcation, will be considered for the Brusselator.

6.3.2.3 The Oregonator with diffusion

Now, we shall write equations for a reaction with diffusion assuming that at the front of a chemical wave $z = 0$ $(z \equiv [\mathrm{Ce}^{4+}])$ which implies that the cerium ion is in the Ce^{3+} state and the first two reactions of the mechanism

(6.13) dominate in the BZ reaction. Upon taking into account diffusion and the condition $z = 0$, equations (6.48) become

$$\partial x/\partial t = k_1 ay - k_2 xy + k_3 ax - 2k_4 x^2 + D\partial^2 x/\partial r^2 \tag{6.140a}$$

$$\partial y/\partial t = -k_1 ay - k_2 xy + D\partial^2 y/\partial r^2 \tag{6.140b}$$

where it was assumed that $D_x = D_y = D$.

Introducing in (6.140) dimensionless variables by means of the substitution

$$t \to [1/(k_3 a)]\, t, \quad r \to [D/(k_3 a)]^{1/2}\, r$$

$$x \to [k_3 a/(2k_4)]\, x, \quad y \to (k_3 a\alpha/k_2)\, y$$

$$L = 2k_4 k_1/(k_2 k_3), \quad M = k_1/k_3, \quad b = k_2/(2k_4) \tag{6.141}$$

where α is a parameter, we obtain

$$\partial x/\partial t = L\alpha y + x(1 - x - \alpha y) + \partial^2 x/\partial r^2 \tag{6.142a}$$

$$\partial y/\partial t = -My - bxy + \partial^2 y/\partial r^2 \tag{6.142b}$$

Note that if we neglect the y variable in (6.142), $y \equiv 0$, then the Fisher–Kolmogorov equation (6.123b), in which $a = 1$, $D = 1$, is obtained for the x variable.

We shall now proceed to the examination of the solution of a travelling wave type. Accordingly, we will substitute in (6.142) for

$$x(t, r) = f(r + vt) \equiv f(\rho) \tag{6.143a}$$

$$y(t, r) = g(r + vt) \equiv g(\rho) \tag{6.143b}$$

obtaining

$$vf' = L\alpha g + f(1 - f - \alpha g) + f'' \tag{6.144a}$$

$$vg' = -Mg - bfg + g'' \tag{6.144b}$$

where prime stands for differentiation with respect to the ρ variable.

The successive substitution, $f = z_1$, $f' = z_2$, $g = z_3$, $g' = z_4$, allows to write equations (6.144) in the form of a system containing only first-order derivatives

$$z_1' = z_2 \tag{6.145a}$$

$$z_2' = -z_1 + vz_2 - L\alpha z_3 + z_1^2 - \alpha z_1 z_3 \tag{6.145b}$$

$$z_3' = z_4 \tag{6.145c}$$

$$z_4' = Mz_3 + vz_4 + bz_1z_3 \tag{6.145d}$$

Inspection of the autonomous system (6.145) will being with finding the stationary points, $\bar{z}_1' = 0, ..., \bar{z}_4' = 0$. The system (6.145) has the following stationary points having the non-negative coordinates z_1, z_3:

$$\bar{z}_1^{(1)} = \bar{z}_2^{(1)} = \bar{z}_3^{(1)} = \bar{z}_4^{(1)} = 0 \tag{6.146a}$$

$$\bar{z}_1^{(2)} = 1, \quad \bar{z}_2^{(2)} = \bar{z}_3^{(2)} = \bar{z}_4^{(2)} = 0 \tag{6.146b}$$

We will linearize the system nearby both the stationary points

$$z_i = \bar{z}_i^{(j)} + \zeta_i, \quad |\zeta_i| \ll 1 \tag{6.147}$$

where $i = 1, ..., 4, j = 1, 2$. In a case of the first stationary state we obtain the following linear system:

$$\zeta_1' = \zeta_2 \tag{6.148a}$$

$$\zeta_2' = -\zeta_1 + v\zeta_2 - L\alpha\zeta_3 \tag{6.148b}$$

$$\zeta_3' = \zeta_4 \tag{6.148c}$$

$$\zeta_4' = M\zeta_3 + v\zeta_4 \tag{6.148d}$$

The characteristic equation (5.9) for the system (6.148)

$$(\lambda^2 - v\lambda - M)(\lambda^2 - v\lambda + 1) = 0 \tag{6.149}$$

has the roots

$$\lambda_{1,2}^{(1)} = 1/2v \pm 1/2(v^2 - 4)^{1/2} \tag{6.150a}$$

$$\lambda_{3,4}^{(1)} = 1/2v \pm 1/2(v^2 + 4M)^{1/2} \tag{6.150b}$$

On the other hand, for the second stationary state we obtain the system

$$\zeta_1' = \zeta_2 \tag{6.151a}$$

$$\zeta_2' = \zeta_1 + v\zeta_2 - (L+1)\alpha\zeta_3 \tag{6.151b}$$

$$\zeta_3' = \zeta_4 \tag{6.151c}$$

$$\zeta_4' = (M + b)\zeta_3 + v\zeta_4 \tag{6.151d}$$

The characteristic equation (5.9) for the system (6.151)

$$[\lambda^2 - v\lambda - (M + b)](\lambda^2 - v\lambda - 1) = 0 \tag{6.152}$$

has the roots

$$\lambda_{1,2}^{(2)} = 1/2v \pm 1/2(v^2 + 4)^{1/2} \tag{6.153a}$$

$$\lambda_{3,4}^{(2)} = 1/2v \pm 1/2[v^2 + 4(M + b)]^{1/2} \tag{6.153b}$$

The roots $\lambda_{1,2}^{(1)}$, $\lambda_{1,2}^{(2)}$, $\lambda_{3,4}^{(2)}$ correspond to unstable nodes for all values of control parameters. In contrast, a pair of the roots $\lambda_{1,2}^{(1)}$ has, for $v = 2$, a sensitive state: when $v = v_{cr} = 2$, $\lambda_1^{(1)} = \lambda_2^{(2)} = 1$.

The roots $\lambda_{1,2}^{(1,2)}$ in this model are identical with the eigenvalues, for $a = D = 1$, in the Fisher–Kolmogorov model. As the second pair of the roots does not lead to a generation of the sensitive states, the nature of a catastrophe for the Oregonator with diffusion is the same as for the Fisher–Kolmogorov model. In other words, the waves with the velocities $v < v_{cr} = 2$ are unstable and, even if generated, they vanish. The fact that equations (6.145) may be reduced to (6.126) follows from the centre manifold theorem: the sensitive state is associated only with the z_1, z_2 variables.

6.3.2.4 The Brusselator with diffusion

We will write a system of equations with diffusion for the mechanism of reaction (6.38). According to principles presented in Section 4.4 the respective equations for a reaction with diffusion are as follows:

$$\partial x/\partial t = a - (b + 1)x + x^2 y + D_x \partial^2 x/\partial r^2 \tag{6.154a}$$

$$\partial y/\partial t = bx - x^2 y + D_y \partial^2 y/\partial r^2 \tag{6.154b}$$

where r corresponds to a linear dimension of a reactor (the case of diffusion along the reactor).

At present, we will examine the problem of the stability of the spatially homogeneous stationary solution, $\bar{x} = a$, $\bar{y} = b/a$, of the Brusselator in terms of the possibile loss of stability of this state and the generation of dissipative structures. Introducing into (6.154) the substitutions

$$x(t, r) = a + \xi(t, r), \quad y(t, r) = b/a + \eta(t, r) \tag{6.155}$$

where $|\xi|$, $|\eta| \ll 1$, we obtain

$$\partial \xi/\partial t = (b - 1)\xi + a^2 \eta + D_x \partial^2 \xi/\partial r^2 \tag{6.156a}$$

$$\partial \eta/\partial t = -b\xi - a^2 \eta + D_y \partial^2 \eta/\partial r^2 \tag{6.156b}$$

Expanding ξ, η in a Fourier series

$$\xi(t, r) = \sum_{n=0}^{\infty} \alpha_n(t) \sin(nr) \tag{6.157a}$$

$$\eta(t, r) = \sum_{n=0}^{\infty} \beta_n(t) \sin(nr) \tag{6.157b}$$

with an assumption that the origin of the reactor corresponds to the coordinate $r = 0$ and its end to the coordinate $r = \pi$ and then substituting for (6.157) in (6.156) we obtain

$$\dot{\alpha}_n = (b - 1)\alpha_n + a^2\beta_n - n^2 D_x \alpha_n \tag{6.158a}$$

$$\dot{\beta}_n = -b\alpha_n - a^2\beta_n - n^2 D_y \beta_n \tag{6.158b}$$

The linear system (6.158) has solutions of the form (5.7), that is $\alpha_n = \alpha_{0,n} \exp(\mu_n t)$, $\beta_n = \beta_{0,n} \exp(\mu_n t)$. The characteristic equation (5.9), corresponding to the system (6.158), thus has the following form:

$$(\mu + n^2 D_x)(\mu + n^2 D_y) - (b - 1)(\mu + n^2 D_y) + a^2(\mu + n^2 D_x) + a^2 = 0 \tag{6.159}$$

or

$$W(n) \equiv [\mu(n)]^2 + B(n)\mu(n) + C(n) = 0 \tag{6.159a'}$$

$$B(n) = a^2 + 1 - b + n^2 D_x + n^2 D_y \tag{6.159b'}$$

$$C(n) = (n^2 D_x)(n^2 D_y) - (b - 1)n^2 D_y + a^2 n^2 D_x + a^2 \tag{6.159c'}$$

where the designations $\mu(n)$, $B(n)$, $C(n)$ were used to emphasize the dependence on n.

When $n = 0$ or $D_x = D_y = 0$ we obtain the characteristic equation of the problem without diffusion

$$\lambda^2 + (a^2 + 1 - b)\lambda + a^2 = 0 \tag{6.160}$$

Consider a situation wherein the stationary state $(a, b/a)$ of the equation without diffusion is stable, that is both the roots of equation (6.160) have negative real parts. It follows from the Routh–Hurwitz criterion (Appendix A5.8) that in this case the control parameters must satisfy the relationships

$$a^2 + 1 - b > 0, \quad a^2 > 0 \tag{6.161}$$

The state $(a, b/a)$ will lose stability if one of the eigenvalues becomes positive. Hence, the sensitive state corresponds to the vanishing of one of the eigenvalues. One of the numbers $\mu_{1,2}$ vanishes when the free term in equation (6.159) is equal to zero

$$C = (n^2 D_x)(n^2 D_y) - (b - 1)n^2 D_y + a^2 n^2 D_x + a^2 = 0 \qquad (6.162)$$

Solving (6.162) for n^2 we obtain

$$(n_{1,2})^2 = (2 D_x D_y)^{-1} \left[A \mp (A^2 - 4a^2 D_x D_y)^{1/2} \right]$$
$$A \equiv (b - 1)D_y - a^2 D_x \qquad (6.163)$$

It follows from the Routh–Hurwitz criterion applied to (6.159′) and from inequality (6.161) that the diffusional instability takes place when $C \leqslant 0$ (since always $B > 0$), that is when the inequality

$$n_1 \leqslant n \leqslant n_2 \qquad (6.164)$$

is valid.

The problem remains to be settled when equation (6.163) has non-negative solutions. The requirement $0 \leqslant (n_{1,2})^2$ is satisfied when the following inequalities

$$0 \leqslant (n_{1,2})^2 \leftrightarrow 1 \leqslant b, \qquad a^2 D_x \leqslant (\sqrt{b} - 1)^2 D_y \qquad (6.165)$$

hold.

The conditions for the occurrence of diffusional instability thus reduce to: inequality (6.161) — the requirement for stability of the spatially homogeneous stationary state $(a, b/a)$ in the absence of diffusion; inequality (6.162), where $n_{1,2}$ are given by (6.163) — the requirement for the loss of stability by the state $(a, b/a)$, and inequality (6.165) — the requirement for correctness of the solution (6.157).

The state $(a, b/a)$ loses stability on crossing by the system, with a continuous variation in the control parameters a, b, D_x, D_y, the sensitive state defined by the equalities in inequalities (6.165). A catastrophe involving the loss of stability of the stationary state (equations without diffusion) due to the effect of diffusion is called the Turing bifurcation.

To be able to determine properties of the dissipative structure being formed, the solutions of the form (6.157) to exact equation (6.154) have to be examined.

A catastrophe of this type is the simplest loss of stability by a system owing to diffusion. It is represented by the sensitive state $\mu_1 = 0$ (μ_2 different from zero). Presently, we will analyse the most sensitive state: that of the largest codimension (of the highest degeneracy), studied by Guckenheimer (we do not assume at this point that inequality (6.161) holds, since we examine a general situation). First, note that the coefficient B in (6.159′) is an increasing function of n. Accordingly, if (6.159') has a pair of purely imaginary roots for $n > 1$, then there exists such l that μ_{n-1} has a positive real part $(B(n) = 0, \ B(n -) < 0)$. Furthermore, $C(n)$ is a second-order polynomial in n^2 with the leading coefficient $D_x D_y > 0$. Thus, if there are zero roots among $\mu_{1,2}(k)$, $\mu_{1,2}(l)$ and $|k - l| > 1$, then there exists such n that $C(n) < 0$, that is $W(n)$ will have a root with a positive real part.

From the above considerations follows a conclusion that if there are no roots of equation (6.159') having a positive real part, then the maximum number of roots lying on the imaginary axis, $\mathrm{Re}(\mu) = 0$, is four. Such a situation is possible if $W(k)$, $W(k + 1)$ have one zero solution for $k > 1$, $W(1)$ has a pair of purely imaginary roots while the remaining roots of equations $W(n)$ have negative real parts. Guckenheimer determined the values of control parameters for which such a most critical situation takes place:

$$D_y = (D_x{}^2 K^2 + 2D_x K)/(1 + D_x K^2) \tag{6.166a}$$

$$a^2 = D_x D_y K^2, \quad K \equiv k(k + 1) \tag{6.166b}$$

$$b = 1 + a^2 + D_x + D_y \tag{6.166c}$$

Equation (6.166c) implies vanishing of the coefficient $B(1)$ in the equation $W(1) = 0$, see (6.159a'), and $B(n) > 0$ for $n > 1$. Next, it follows from the equation for $C(n)$ that if $C(k) = C(k + 1) = 0$, then $C(n) \geqslant 0$ for all n. From the equations $C(k) = C(k + 1) = 0$ we obtain equations (6.166a), (6.166b).

Exact properties of such a state have not been found. Guckenheimer studied the properties of the system nearby a less degenerate stationary state in which the polynomial $W(k)$ has one zero solution for $k > 1$, $W(1)$ has a pair of purely imaginary roots while the remaining roots of the equations $W(n)$ have negative real parts. Then, the control parameters satisfy the requirements

$$a^2 = D_y k^2 \left[(D_x + D_y - D_y k^2)/(1 + D_x k^2 - D_y k^2) \right] \tag{6.167a}$$

$$b = 1 + a^2 + D_x + D_y \tag{6.167b}$$

$$D_y k^2 \left[(D_x + D_y - D_y k^2)/(1 + D_x k^2 - D_y k^2) \right] \left[1 + (k \pm 1)^2 (D_x - D_y) \right]$$
$$- (k \pm 1)^2 D_y (D_x + D_y) + (k \pm 1)^4 D_x D_y > 0 \tag{6.167c}$$

Guckenheimer reduced equations (6.156), for the parameters fulfilling the conditions (6.167), by substituting into them expansions of the type (6.157) containing only the terms corresponding to sensitive states

$$\xi(t,r) = \alpha_1(t)\sin(r) + \alpha_k(t)\sin(kr) \tag{6.168a}$$

$$\eta(t,r) = \beta_1(t)\sin(r) + \beta_k(t)\sin(kr) \tag{6.168b}$$

since it follows from the centre manifold theorem that such a simplification allows us to describe the possible catastrophes. The substitution enables the reduction of (6.156) to a certain autonomous system of two equations. A theoretical analysis and computer calculations indicate the existence of chaotic solutions.

6.3.3 Chemical chaos

At the end of this chapter we will now briefly discuss a theoretical approach to the description of chaotic processes encountered in chemical kinetics, see Section 1.3 and Sections 6.2.2.4, 6.3.2.4. In Section 6.2.2.4 we described the method of generation and physical meaning of a chaotic state of the Belousov–Zhabotinskii reaction carried out in a flow reactor.

The chaotic dynamics can be conveniently followed by monitoring a concentration of just one reagent, for example thr Br^- ion (using a suitable ion-selective electrode or colorimetrically). The measurement of the Br^- concentration at constant time intervals τ is most straightforward. The consecutively measured concentrations will be

$$[Br^-](t_1), \quad [Br^-](t_1 + \tau), \quad [Br^-](t_1 + 2\tau), \dots$$

where t_1 is the time of the first measurement. The experimental results can be summarized as follows. For a small flow rate v decaying oscillations of a sequence of the measured $[Br^-](t_1 + n\tau)$ values are observed. On increasing v periodical oscillations take place and a further increase in

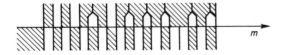

Fig. 102. Feigenbaum cascade in the Oregonator with flow. Reprinted with permission from: R.H. Simoyi, A. Wolf and H.L. Swinney, *Physical Review Letters*, **49** (1982), 245.

v gives rise to a complication in the nature of oscillations. For increasing v oscillations having a period larger by a factor of 2, 2^2, 2^3 than that of the initial oscillations have been observed. For still larger v values chaotic oscillations are observed, see Fig. 102. It should be pointed out that in the entire range of variability of the parameter v the flow of reagents through the reactor was laminar.

This type of transition to a chaotic state is called the Feigenbaum cascade. Such a scenario of generation of chaos involves consecutive losses of stability by successive orbits, see Fig. 103.

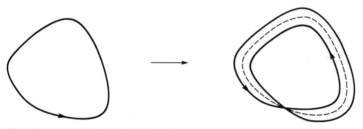

Fig. 103. Period doubling bifurcation.

The measurements of x_i reveal that the observed phenomena can be modelled by means of the recurrent equation

$$x_{i+1} = 4bx_i(1 - x_i) \tag{6.169a}$$

in which the parameter b is a function of the flow rate of reagents through the reactor since, as we have shown in Section 3.6, a transition to a chaotic state by way of the Feigenbaum cascade takes place in the system (6.169a).

The experimental phenomena observed in the Belousov–Zhabotinskii reaction, such a doubling of the oscillation period, chaotic oscillations or alternate periodical and chaotic oscillations, can be modelled still more exactly by the recurrent equation

$$x_{i+1} = f(x_i) \tag{6.169b}$$

in which the function $f(x)$ is properly selected (such a function resembles the function xe^{-x}).

A chaotic nature of oscillations observed in the BZ reaction can be visualized by drawing a graph in three dimensions of the coordinates $x(t)$, $x(t + \tau)$, $x(t + 2\tau)$, where this time $x = [Ce^{4+}]$, see Fig. 104.

The trajectories in the figure have a chaotic, turbulent character, typical

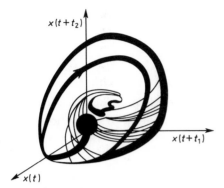

Fig. 104. Chemical chaos in the Belousov–Zhabotinskii reaction. Reprinted with permission from: C. Vidal, page 49, Springer Series in Synergetics, H. Haken (Ed.), Vol. 12. *Nonlinear Phenomena in Chemical Dynamics.*

for strange attractors, see for example the Lorenz system described in Chapter 5. Hence, it can be presumed that the kinetic equations of the Oregonator, taking into account the flow of reagents through the reactor,

$$\dot{x} = k_1 ay - k_2 xy + k_3 ax - 2k_4 x^2 + 1/T(x_0 - x) \qquad (6.170\text{a})$$

$$\dot{y} = -k_1 ay - k_2 xy + hk_5 z + 1/T(y_0 - y) \qquad (6.170\text{b})$$

$$\dot{z} = 2k_3 ax - k_5 z + 1/T(z_0 - z) \qquad (6.170\text{c})$$

in which T is the retention time inversely proportional to the flow rate v of reagents through the reactor, see Section 6.2.4, should have for suitable T values the solutions of a chaotic nature. Indeed, this conclusion has been supported by computer analysis.

To end with, let us mention that the dissipative structures of a turbulent nature can be modelled by accounting in the Oregonator with flow equations (6.170) for the possibility of diffusion of reagents. The structures of this type have been observed experimentally. They can be modelled by means of equations of the type (6.120′) in one state variable x

$$\partial x/\partial t = P(x, r) + \partial/\partial r [D(x, r)\partial x/\partial r] \qquad (6.171\text{a})$$

For example, the equation obtained for properly selected functions P, D,

$$\partial x/\partial t = Ax - x\partial x/\partial r + D\partial^2 x/\partial r^2 + U_0 \sin(Br) \qquad (6.171\text{b})$$

and corresponding to a reaction with diffusion in the presence of a periodical disturbance proportional to the constant U_0, represents chaotic spatial structures.

6.4 CATASTROPHES ASSOCIATED WITH THE CHANGES IN DISTRIBUTION OF MOLECULAR ELECTRON DENSITY

6.4.1 Introduction

Catastrophe theory also finds use in the description of chemical reactions in terms of individual molecules participating in a reaction. The quantity examined is the molecular charge distribution ρ, which is defined by the following integral:

$$\rho\left(\mathbf{r}_i, \mathbf{R}, t\right) = N \sum_{\text{spin}} \int \prod_{j=1} \mathrm{d}^3 \mathbf{r}_j \, \Psi^*\left(\mathbf{r}, \mathbf{R}, t\right) \Psi\left((\mathbf{r}, \mathbf{R}, t)\right) \tag{6.172}$$

where Ψ is the wave function of a molecule, N is the number of electrons, the vector $\mathbf{r} = (\mathbf{r}_1, ..., \mathbf{r}_N)$, where $\mathbf{r}_i = (x_i, y_i, z_i)$ represents the positions of electrons and the vector $\mathbf{R}$ describes a fixed configuration of atomic nuclei.

Control parameters are the positions of atomic nuclei $\mathbf{R}$, a function of state is the wave function Ψ satisfying the time-dependent Schrödinger equation — an equation of state. The problem of determining the dependence of the integral (6.172) on parameters is related to the problem associated with the investigation of oscillatory integrals occurring in optics and in diffraction problems, see Section 3.4.

6.4.2 Description of the method of analysis of electron density distributions

Topological properties of the molecular distribution of electron density, given by (6.172), are closely related to critical points of the function ρ, i.e. the points at which

$$\nabla_r \rho\left(\mathbf{r}_i, \mathbf{R}, t\right) = 0 \tag{6.173}$$

Below we shall try to explain the relation between the equality of the type (6.173) and the topology of charge distribution ρ. Consider a certain region V having an edge S in the space of electron coordinates $\mathbf{r}_i$. Let the vector $\mathbf{n}$ be perpendicular to S and directed outward the region V. The law of conservation of charge is of the form

$$\partial/\partial t \rho + \nabla \cdot \mathbf{j} = 0 \tag{6.174}$$

where $\mathbf{j}$ is the current flowing through the surface. If we demand conservation of the charge inside the volume V

$$\partial/\partial t\, Q \equiv \partial/\partial t \int_v \rho \, \mathrm{d}v \qquad (6.175)$$

then we obtain from (6.175), (6.174)

$$0 = \int_v \nabla \cdot \mathbf{j} \, \mathrm{d}v = \int_s \mathbf{jn} \, \mathrm{d}s \qquad (6.176)$$

where the Gauss theorem was applied. Equation (6.176) will hold if the expression representing current flowing perpendicularly to the surface S vanishes on S:

$$\mathbf{j}(\mathbf{r}) \cdot \mathbf{n} = 0, \quad \mathbf{r} \in S \qquad (6.177)$$

The current $\mathbf{j}$ is proportional to the gradient of charge distribution, $\mathbf{j} \sim \nabla \rho$; (6.177) may thus be written in the following form:

$$[\nabla_r \rho(\mathbf{r}_i, \mathbf{R}, t)] \cdot \mathbf{n} = 0, \quad \mathbf{r}_i \in S \qquad (6.178)$$

Hence, if there exists a closed surface S such that equation (6.178) is valid, then the charge contained in the volume V, bounded by this surface, is conserved.

Note that from the mathematical standpoint the condition (6.178) implies that $\nabla_r \rho$ is parallel to the tangent to the surface S. Instead of examining the expression occurring in (6.178), the density Laplacian can be analysed, since substituting into the volume integral in (6.176) for $\mathbf{i} \sim \nabla \rho$, the condition (6.178) may be replaced by the requirement

$$\Delta[\rho(\mathbf{r}_i, \mathbf{R}, t)] = 0, \quad \mathbf{r}_i \in S \qquad (6.179)$$

Consider now the charge distribution ρ given by (6.172) and corresponding to a specified molecule. The closed surface S, on which the requirement (6.178) holds, divide the molecule and the surrounding space into the regions V, between which there is no charge flow. It turns out that every such a region V contains one atom of the molecule.

Examine now a certain chemical reaction which will be symbolically written as a transformation of the reactant A into the product B:

$$A \rightarrow B \qquad (6.180)$$

In the course of the reaction the charge density is transformed as follows

$$\rho(\mathbf{r}_i, \mathbf{R}_A, t) \rightarrow \rho(\mathbf{r}_i, \mathbf{R}, t) \rightarrow \rho(\mathbf{r}_i, \mathbf{R}_B, t) \qquad (6.181)$$

where $\mathbf{R}_A$ stands for the position of nuclei in a molecule of the reactant (reactants), A, $\mathbf{R}_B$ is the position of atomic nuclei in a molecule of the product (products) B, $\mathbf{R}$ is the intermediate position of atomic nuclei on the path of the reaction. The vector $\mathbf{R}$ plays the role of control parameters while $\rho(\mathbf{r}_i, \mathbf{R}, t)$ is a state function.

The charge distribution $\rho(\mathbf{r}_i, \mathbf{R}, t)$ on the curve of the reaction may be calculated by quantum mechanics methods; hence we shall assume that for the reaction (6.181) this distribution is known. For each $\mathbf{R}$ value on the path of the reaction we can determine the charge distribution gradient.

The closed surfaces S, on which the requirement (6.179) is satisfied, allow us to divide a molecule being present on the path of the reaction, into atoms. We say that a catastrophe took place if there occurs a qualitative change in the charge density distribution ρ and in its gradient $\nabla_r\rho$ on crossing by the control parameters $\mathbf{R}$ the critical position $\mathbf{R}_{cr}$. Such a catastrophe is connected with a qualitative change in the division of a molecule being on the reaction path into atoms; hence it reflects

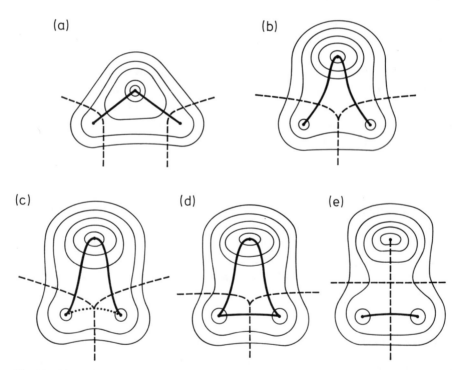

Fig. 105. Plots of electron density for the reaction $H_2O \rightarrow H_2 + O$. Reprinted with permission from: R.F.W. Bader, T.T. Nguyen-Dang and Y. Tal, *Reports on Progress in Physics*, **44** (1981), 893.

a qualitative alteration within an assembly of reacting molecules — that is, the occurrence of a chemical reaction.

The above considerations will be illustrated by the results of computations for the reaction

$$H_2O \rightarrow H_2 + 0 \tag{6.182}$$

The results of computations are shown in Fig. 105.

In the course of the reaction hydrogen atoms move away from an oxygen atom and approach each other. The plots of electron density with the marked surfaces S dividing the water molecule into atomic regions permit to detect the sensitive state corresponding to a catastrophe — the occurrence of a reaction. Figure 105c depicts the sensitive state on crossing which the system of atoms composing the H_2O molecule should rather be written as $\{O + H_2\}$.

In general, sensitive states according to elementary catastrophe theory correspond to such values of the control parameters $\mathbf{R}$ that the critical point of the function ρ, fulfilling the condition (6.178), is a degenerate point. In other words, at such a point the following requirement holds:

$$\det \left| \partial^2 \rho / \partial x_i \partial x_j \right| = 0 \tag{6.183}$$

Critical points of the density function ρ can be classified in accordance with elementary catastrophe theory. For instance, Bader has demonstrated that in the case of reaction (6.182) the catastrophe of an elliptic umbilic type (D_4^-) takes place.

6.5 CONCLUSION

In the first quarter of the twentieth century a number of fundamental experimental discoveries and theoretical predictions have been made. In 1905 Liesegang discovered the phenomenon of generation of spatially periodical structures, the so-called Liesegang rings, on precipitating certain compounds from homogeneous solutions. In 1910 Lotka predicted theoretically the possibility of occurrence of oscillations of reagent concentrations in a chemical reaction. Some time later Morgan observed a pulsatory evolution of CO from aqueous HCOOH solutions and Brag revealed periodicity in the reaction of H_2O_2 decomposition in the presence of IO_3^-/I_2. A variety of chemical and biochemical systems exhibiting time and spatial periodical properties are now known. Ever since the first paper by

Lotka a considerable progress in examining and understanding very diverse non-linear phenomena observed in chemical reactions has been achieved. In addition to periodical phenomena, multistability, diversified wave phenomena, time and spatial chemical chaos have also been observed. Owing to Prigogine's works it has been understood that such phenomena should be sought far from the state of thermodynamic equilibrium (stationary state). The states appearing far from the equilibrium may in some cases be enhanced and stabilized with the generation of a new, stable dynamical state. Processes of this type may be readily investigated in flow reactors in which a natural measure of deviation from the equilibrium state is the retention time. The ease of generation of very diversified states in flow reactors opens new outlooks in a rapidly developing area of nonlinear phenomena.

On the other hand, the abundance of experimental material stimulates an evolution of the theories explaining non-linear phenomena. For example, as shown above, the transition in a chemical reaction from the stationary state to the state of periodical oscillations, the so-called Hopf bifurcation, is a certain elementary catastrophe. The transition in a chemical reaction to the chaotic state may be explained in terms of catastrophes associated with a loss of stability of a certain iterative process or by using the notion of a strange attractor (anyway, it turns out that both the systems are closely related). The equations of a chemical reaction with diffusion have been extensively studied lately. Based on the progress being made in this area, further interesting achievements in theory may be anticipated, particularly for the phenomena associated with catastrophes — the loss of stability by a non-linear system.

Bibliographical Remarks

The Belousov–Zhabotinskii reaction has been described and analysed in books by Zhabotinskii and Tyson, as well as by Murray. Much information on the models of chemical reactions is provided in books by Prigogine and coworkers and by Chernavskii, Romanovskii and Stepanova. Chemical chaos is described in a paper by Agladze and Krinsky, Simoyi *et al.* and in collections of papers edited by Kadomcev, Vidal and Pacault, as well as a book by Bergé. The collections also contain a number of interesting papers dealing with other non-linear phenomena, such as multistability, dissipative structures and chemical waves. A book edited by Field and Burger contains a comprehensive set of papers dealing with investigations on various dynamical states of chemical kinetics.

Applications of catastrophe theory to a description of chemical reactions from the standpoint of individual molecules taking part in a reaction are described in papers by Bader and by Connor.

References

K. I. Agladze and V. I. Kinsky, "Multi-armed vortices in an active chemical medium", *Nature*, **296**, 424 (1982).

K. I. Agladze, A. V. Panfilov and A. N. Rudenko, "Nonstationary rotation of spiral waves: three dimensional effect", *Physica*, **D29**, 409 (1988).

L. Arnol'd and R. Lefever (Eds.), *Stochastic Nonlinear Systems in Physics, Chemistry and Biology*, Springer Series in Synergetics, Vol. 8, Springer-Verlag, Berlin–Heidelberg–New York, 1981.

R. F. W. Bader and T. T. Nguyen-Dang, *Adv. Quant. Chem.*, **14**, 63 (1981).

R. F. W. Bader, T. T. Nguyen-Dang and Y. Tal, "The topological theory of molecular structure", *Repts. Progr. Phys.*, **44**, 893 (1981).

R. J. Bagley, G. Mayer-Kress and J. D. Farmer, "Mode locking, the Belousov–Zhabotinsky reaction and one-dimensional mappings", *Phys Lett.*, **A114**, 419 (1986).

P. Bergé, Y. Pomeau and Ch. Vidal, *L'ordre dans le chaos*, Hermann, Paris, 1984.

P. P. Belousov, "Periodicheskaya rieakcya y yeyo miechanism", *Sb. Ref. Radiat. Med.* 1958. Moskva 145 (1959).

V. N. Biktashev, "Evolution of twist of an autowave vortex", *Physica*, **D36**, 167 (1989).

J. N. L. Connor, "Catastrophes and molecular collisions", *Mol. Phys.*, **31**, 33 (1976).

C. Cosubr and A. C. Lazer, "The Volterra–Lotka model with diffusion", *SIAM J. Appl. Math.*, **44**, 1112 (1984).

J. P. Dockerey, J. P. Keener and J. J. Tyson, "Dispersion of travelling waves in the Belousov–Zhabotinskii reaction", *Physica*, **D30**, 177 (1988).

W. Ebeling, *Strukturbildung bei Irreversibelen Prozessen*, BSB B. G. Teubner Verlagsgesellschaft, Rostock, 1976.

I. R. Epstein and K. Kustin, *Design of Inorganic Chemical Oscillators, Structure and Bonding*, Vol. 56, Springer-Verlag, Berlin–Heidelberg–New York–Tokyo, 1984.

D. Feinn and P. Ortoleva, "Catastrophe and propagation in chemical reactions", *J. Chem. Phys.*, **67**, 2119 (1977).

R. J. Field and M. Burger (Eds.), *Oscillations and Travelling Waves in Chemical Systems*, John Wiley and Sons, New York–Chichester–Brisbane–Toronto–Singapore, 1985.

R. J. Field, E. Körös and R. M. Noyes, "Oscillations in chemical systems, Part 2. Thorough analysis of temporal oscillations in the $Ce-BrO_3^-$ — malonic acid system", *J. Am. Chem. Soc.*, **94**, 8649 (1972).

P. C. Fife, "Mathematical aspects of reacting and diffusing systems", *Lecture Notes in Biomath.*, **28**, 1 (1979).

P. C. Fife, "The Belousov–Zhabotinski reagent", *J. Stat. Phys.*, **39**, 687 (1985).

P. Glansdorf and I. Prigogine, *Structure, stabilité et fluctuations*, Masson et Cie, Paris, 1971.

J. Guckenheimer, „Multiple bifurcation problems of codimension two", *SIAM J. Math. Anal.*, **15**, 1 (1984).

J. Guckenheimer, "Multiple bifurcation problems for chemical reactors", *Physica*, **20D**, 1 (1986).

H. Haken, *Synergetics. An Introduction. Nonequilibrium Phase Transitions and Self-Organization in Physics, Chemistry and Biology*, Springer-Verlag, Berlin–Heidelberg–New York, 1978.

H. Haken, *Synergetics*, Springer-Verlag, Berlin–Heidelberg–New York, 1978.

B. D. Hassard, N. D. Kazarinoff and Y.-H. Wan, *Theory and Applications of Hopf Bifurcation*, Cambridge University Press, Cambridge, 1981.

B. B. Kadomcev (Ed.), *Syniergetika* (a collection of papers), Mir, Moskva, 1984.

J. P. Keener, "The dynamics of three-dimensional scroll waves in excitable media", *Physica*, **D31**, 269 (1988).

J. P. Keener, "Knotted scroll wave filamentin excitable media", *Physica*, **D34**, 378 (1989).

L. Kuhnert, L. Pohlman and H.-J. Krug, "Chemical wave propagation with a chemically induced hydrodynamical instability", *Physica*, **D29**, 416 (1988).

A. J. Lotka, "Contribution to the theory of periodic reactions", *J. Phys. Chem.*, **14**, 271 (1910).

A. J. Lotka, *Elements of Physical Biology*, Williams and Williams, Baltimore, 1925.

J. Masełko, M. Alamgir and I. R. Epstein, "Bifurcation analysis of a system of coupled chemical oscillators: bromate–chlorite–iodite". *Physica*, **19D**, 153 (1986).

J. Masełko and K. Showalter, "Chemical waves on spherical surfaces", *Nature*, **339**, 609 (1989).

T. Matsumoto, L. O. Chua and M. Komuro, "Birth and death of the double scroll", *Physica*, **D24**, 110 (1987).

S. C. Müller, T. Plesser and B. Hess, "Two dimensional spectrophotometry of spiral wave propagation in the Belousov–Zhabotinskii reaction", *Physica*, **D24**, 71 (1987).

J. D. Murray, *Lectures on Nonlinear–Differential–Equation Models in Biology*, Clarendon Press, Oxford, 1977.

G. Nicolis, "Chemical instabilities", *Physica*, **17D**, 345 (1985).

G. Nicolis, "Dissipative systems", *Repts. Progr. Phys.*, **49**, 873 (1986).

G. Nicolis and F. Baras, "Chemical instabilities", NATO Adv. Science Institutes Series C. Math. and Physical Sciences 120.

G. Nicolis and I. Prigogine, *Proc. Nat. Acad. Sci. (USA)*, **68**, 2102 (1971).

G. Nicolis and I. Prigogine, *Self-Organization in Nonequilibrium Systems*, John Wiley and Sons, New York, 1977.

A. V. Panfilov, R. R. Aliev and A. V. Mushinsky, "An integral invariant for scroll rings in a reaction–diffusion system". *Physica*, **D36**, 181 (1989).

J. M. Romanovskii, N. W. Stepanova and D. S. Chernavskii, *Mathematical Biophysics* (in Russian), Nauka, Moskva, 1984.

A. Schmitz, K. A. Graziani and J. L. Hudson, "Experimental evidence of chaotic states in the Belousov–Zhabotinskii reaction", *J. Chem. Phys.*, **67**, 3040 (1977).

R. H. Simoyi, A. Wolf and H. L. Swinney, "One dimensional dynamics in a multi component chemical reaction", *Phys. Rev. Lett.*, **49**, 245 (1982).

A. M. Turing, "The chemical basis of the morphogenesis", *Phil. Trans. Roy. Soc. (London)*, **B237**, 37 (1952).

J. J. Tyson, "The Belousov–Zhabotinskii reaction", *Lect. Notes in Biomath.*, **10**, 1 (1976).

C. Vidal and A. Pacault (Eds.), *Chemical Instabilities*, Springer Tract in Synergetics, Springer--Verlag, Berlin–Heidelberg–New York, 1984.

C. Vidal and A. Pacault (Eds.), *Nonlinear Phenomena in Chemical Dynamics*, Springer Tracts in Synergetics, Vol. 12, Springer-Verlag, Berlin–Heidelberg–New York, 1981.

A. T. Winfree, *The Geometry of Biological Time*, Springer-Verlag, New York–Heidelberg--Berlin, 1980.

A. T. Winfree, *Organizing Centres for Chemical Waves in Two and Three Dimensions. Oscillations and Travelling Waves in Chemical Systems*, R. J. Field and M. Burger (Eds.), John Wiley and Sons, New York, 1985.

A. T. Winfree and S. H. Strogatz, "Singular filaments organize chemical waves in three dimensions. I. Simple waves", *Physica*, **8D**, 35 (1983); "II. Twisted waves", *ibid.*, **9D**, 65 (1983); "III. Knotted waves", *ibid.*, **10D**, 333 (1983).

A. M. Zhabotinskii, *Koncentracionnye awtokolebaniya*, Nauka, Moskva, 1974.

Index